HOW TO RESTORE
Classic JOHN DEERE
TRACTORS

The Ultimate Do-It-Yourself Guide to
Rebuilding and Restoring Deere Two-Cylinder Tractors

Tharran E. Gaines

Voyageur Press

Dedication

This book is dedicated to my wife, Barb, and our grown children, Michael and Michelle. Thanks for never trying to take the country out of me.

Text © 2003 by Tharran E. Gaines
Photography © 2003 by Tharran E. Gaines, except where noted

All rights reserved. No part of this work may be reproduced or used in any form by any means—graphic, electronic, or mechanical, including photocopying, recording, taping, or any information storage and retrieval system—without written permission of the publisher.

Edited by Amy Rost-Holtz
Designed by Maria Friedrich
Printed in China

04 05 06 07   5 4 3 2

Library of Congress Cataloging-in-Publication Data

Gaines, Tharran E., 1950-
  How to restore classic John Deere tractors : the ultimate do-it-yourself guide to rebuilding and restoring deere two-cylinder tractors / Tharran E. Gaines.
    p. cm.
  ISBN 0-89658-601-4
  1. John Deere tractors—Maintenance and repair—Handbooks, manuals, etc. I. Title.
  TL233.5.J63 G35 2003
  629.28'752—dc21

2002151943

Distributed in Canada by Raincoast Books
9050 Shaughnessy Street, Vancouver, B.C. V6P 6E5

Published by Voyageur Press, Inc.
123 North Second Street, P.O. Box 338
Stillwater, MN 55082 U.S.A.
651-430-2210, fax 651-430-2211
books@voyageurpress.com
www.voyageurpress.com

Educators, fundraisers, premium and gift buyers, publicists, and marketing managers: Looking for creative products and new sales ideas? Voyageur Press books are available at special discounts when purchased in quantities, and special editions can be created to your specifications. For details contact the marketing department at 800-888-9653.

Legal Notice
This is not an official publication of Deere & Company. The name John Deere, as well as certain names, model designations, and logo designs, are the property of Deere & Company Inc. We use them for identification purposes only. Neither the authors, photographers, publisher, nor this book are in any way affiliated with Deere & Company.

On the frontispiece:
Top: Like its sibling, the Model B, John Deere's Model A was available in several specialized variations, including the AR, AO, and AOS. (Photograph by Hans Halberstadt)

Bottom left: Some restorers get more enjoyment out of restoring tractors than owning them. They simply restore them with the idea of reselling them when they are finished. Either way, it can be a dirty job at times.

Bottom right: A cutoff tool or band saw will come in handy if your restoration requires much metal work.

On the title page:
Top left: When you finish with that last coat of paint, you'll agree that even the dirtiest, most labor intensive jobs were worth the effort.

Top right: Roy Ritter, who retired after forty-five years of working as a John Deere mechanic, currently rebuilds John Deere two-cylinder injection pumps and injectors for restorers throughout the United States.

Bottom: A collection of styled and unstyled tractors await restoration outside the John Deere Collectors Center in Moline, Illinois.

Acknowledgments

One of the first questions I get asked when people learn that I've authored a book on tractor restoration is "Do you restore antique tractors yourself?" So I'll be honest with you right up front. No, I don't have any classic farm tractors myself.

What I do have is more than twenty-five years of experience writing about farm equipment and more than ten years experience as a technical writer. For years, I wrote owner's manuals, repair manuals, and assembly instructions for companies like Sundstrand, Hesston, Winnebago, Kinze, and Best Way crop sprayers. My job was to glean information from a variety of sources in marketing, engineering, product service, and the test lab, and turn it into text that could be used by the customer to assemble, service, or repair the machine he or she had purchased from my employer.

I have approached tractor restoration in much the same way. I figure the people I have talked to and photographed while preparing this manuscript have forgotten more than I could ever learn about tractor restoration. So, in reality, this book is their story, not mine.

With that in mind, I want to express my appreciation to a number of people, starting with my wife, Barb, who has not only blessed me with her encouragement and patience, but has spent several hours proofreading the copy and the photo captions for typographical and grammatical errors.

I also want to thank all of the John Deere enthusiasts and tractor restorers who have devoted their time and knowledge to this project. Without their help, I would not have been able to produce this book. Among them are Roy Ritter and Estel Theis. Both are John Deere two-cylinder tractor enthusiasts who live within fifteen minutes of my home in Savannah, Missouri. Consequently, it was easy for me to run out to their farms a couple times a week and take a few more photos as they completed another step in the restoration process.

Introduced in 1952, the Model 60 was an updated and enhanced version of Deere & Company's popular Model A. (Photograph by Hans Halberstadt)

Having spent more than forty-five years as a John Deere mechanic, Roy is a walking encyclopedia on two-cylinder John Deere tractors. Even today, you can often find him in his farm shop, working on diesel pumps and injectors for people from all over the United States.

Estel Theis, on the other hand, is a full-time farmer who likes to spend his winters in the shop bringing old John Deere two-cylinder tractors back to life. Even though he has nearly two dozen restored tractors in his collection, there's always room for one more. And every year, he manages to complete one or two new models.

Other John Deere and antique tractor enthusiasts who assisted with this book are Rex Miller, a retired Avenue City, Missouri, farmer who repairs tractor magnetos in his spare time; Bill Anderson, a full-time vintage tractor restorer from Superior, Nebraska; Brock Ekhoff, a John Deere salesman and antique tractor enthusiast from Clay Center, Nebraska; Dennis Funk, a farmer and John Deere enthusiast from Hillsboro, Kansas; Ed Hoyt, a retired truck driver and Air Guardsman from St. Joseph, Missouri, who has restored three John Deere tractors since retiring. A few more worth mentioning are Gene Tencza, a John Deere enthusiast from Orange, Massachusetts; Bill Batt, who owns a two-cylinder tractor and salvage business near Garden City, Missouri; Jeff Gravert, a full-time tractor restorer from Central City, Nebraska; and Harold Hatfield, a 20 Series collector from Salisbury, Missouri.

I also owe a great deal of thanks to the staff at Moline Tractor and Plow Company in Moline, Illinois, for their help with tips, research, and photographs. They include Jeff McManus, a retired general manager who now serves as senior consultant; Scott Carlson, service manager; and Tony Bormann and Brian Holst, who manage the parts counter.

Two more people who were of tremendous assistance were Chris and Kim Pratt, who write and edit the *Yesterday's Tractors* on-line magazine. Due to my limited knowledge on some restoration subjects, I drew heavily on information available on their web site, which I would urge you to check out for yourself.

Acknowledgments / 7

While there is a lot of work involved in a tractor restoration, there's nothing like firing up an engine you overhauled yourself and driving a finished tractor for the first time.

It is not only an excellent source of tips and advice on restoration, but it also offers a variety of parts, kits, and manuals.

Others who assisted with photos or answers include the staff at O'Reilly Auto Parts and Bill Briner and his staff at Bill's Auto Electric in Savannah, Missouri. I'm additionally indebted to Hermie Bentrup, owner of Auto Body Color, Inc., in St. Joseph, Missouri, and B. J. Rosmolen, owner of BJ's Auto Collision and Restoration, also in St. Joseph, for their help and advice on paint and painting. Meanwhile, Travis Jorde, owner of Jorde's Decals in Rochester, Minnesota, provided a tremendous amount of advice on decal application.

I want to thank the crew at Custom Color and their E6 City division, in Kansas City, Missouri, for their careful handling and processing of several hundred color slides.

Most of all, though, I want to thank Amy Rost-Holtz with Voyageur Press for her editorial guidance and direction on this project. She and her boss, Michael Dregni, have been a pleasure to work with over the past couple of years.

Estel Theis is as meticulous about the lights as he is about any other tractor component.

The reasonable cost and simplicity puts John Deere two-cylinder tractor restoration within reach of amateur and professional restorers alike.

Contents

Introduction	**Before You Begin 10**	**Chapter 6**	**Engine Repair and Rebuilding 60**

Introduction **Before You Begin 10**
  Know Your Deere 10
  Always Put Safety First 11

Chapter 1 **Two-Cylinder History 12**
  The Beginning of a Legacy 13
  The Poppin' Johnny Sound 15
  Thirty Years of D Models 16
  More Models Followed 18
  Deere Designs a Winner 19
  New York Styling Comes to the Midwest 21
  A Deere of a Different Kind 22
  Numbers Replaced Letters 24

Chapter 2 **Shopping for a Tractor 30**
  Search the Records 33
  How to Tell Them Apart 35
  Buying the Right Tractor 36
  Negotiating a Price 36

Chapter 3 **Setting Up Shop 40**
  Basic Tools 40
  Specialized Tools 42
  Purchase a Good Shop Manual 45

Chapter 4 **Getting Started 46**
  Take Your Time 46
  Establish Your Goals 46
  Clean It Up 48
  Disassembly 49
  Finding Replacement Parts 50
  Removing Broken or Damaged Bolts 52
  Shaft Repair 53

Chapter 5 **Troubleshooting 54**
  Evaluating a Tractor That Runs 55
  Simple Fixes 57
  Compression Testing 58

Chapter 6 **Engine Repair and Rebuilding 60**
  Freeing a Stuck Engine 62
  Ring Job or Complete Overhaul? 64
  Engine Block Preparation 64
  Cylinder Head and Piston Removal 64
  Piston Ring Replacement 69
  Main and Connecting Rod Bearings 74
  Camshaft Overhaul 77
  Valvetrain Overhaul 80
  Pistons and Bores 86
  Rebuilding the Engine 87
  Oil Pump Restoration 92
  Pony Engines 93

Chapter 7 **Clutch, Transmission, and PTO 94**
  Transmission Inspection and Repair 95
  Clutch Inspection and Rebuilding 97
  Power Takeoff Repair 101

Chapter 8 **Final Drive and Brakes 102**
  Differentials and Final Drives 103
  Differential and Final Drive Inspection and Repair 105
  Axle Shafts 106
  Brake Restoration 108
  Brake Adjustment 109

Chapter 9 **Front Axle and Steering 110**
  Steering Through Deere History 110
  Steering Configurations 112
  Front Axle Repair 113
  Power Steering 115
  Steering Wheel Repair 117

Chapter 10	**Tires, Rims, and Wheels** 118	**Chapter 15**	**Sheet Metal** 172

Chapter 10   **Tires, Rims, and Wheels** 118
    Rear Wheel Removal 119
    Tire Repair and Restoration 120
    Wheel and Rim Restoration 121

Chapter 11   **Hydraulic System** 126
    Basic Principles 128
    Contamination 129
    Troubleshooting 130
    Hydraulic Seals 131

Chapter 12   **Electrical System** 132
    Magneto Systems 132
    Magneto Inspection and Service 136
    Timing the Magneto 136
    Distributor Inspection and Repair 138
    Coils 140
    Generators and Voltage Regulators 140
    Starters 143
    Wiring 144
    Spark Plug Wires 146
    Lights 146
    Gauges 147

Chapter 13   **Fuel System** 148
    Fuel Tank 148
    Fuel Hose Inspection 151
    Carburetor Repair 151
    Carburetor Rebuilding 153
    Carburetor Reassembly 155
    Diesel Systems 156
    Oil-Bath Air Filters 158
    Manifold Inspection and Repair 159
    Governor Overhaul 160

Chapter 14   **Cooling System** 162
    Cooling System Inspection and Repair 162
    Shutters and Curtains 165
    Radiator Cap 166
    Fan 166
    Hoses 168
    Water Pump 169
    Thermostat 171
    Belts 171

Chapter 15   **Sheet Metal** 172
    Paint Removal 173
    Repairing Sheet Metal 176

Chapter 16   **Paint** 182
    The Primer Coat 182
    Choosing the Right Primer Type 183
    The Color Coat 185
    Selecting the Right Paint Type 185
    Painting Equipment 187
    Applying the Paint 188
    Finishing the Axles 197

Chapter 17   **Decals, Name Plates, and Serial Number Plates** 198
    Researching Decal Originality and Placement 199
    Tools and Supplies 202
    Surface Preparation 202
    Decal Application 202
    Emblems and Name Plates 206
    Serial Number Plates 206

Chapter 18   **The Fruits of Your Labor** 208
    Antique Tractor Shows 210
    Farm Equipment Demonstrations 213
    Tractor Games 214
    Antique Tractor Pulls 214

Appendix   **Sources for Parts, Rebuilding, and Repairs** 218
    Restoration Equipment and Supplies 221
    Publications and Clubs 221

Introduction

Before You Begin

Know Your Deere

The various restoration procedures outlined or described in this book are often stated as being applicable to "Waterloo-built" or "Dubuque-built" tractors, rather than individual models. Waterloo-built tractors are those built in Waterloo, Iowa, where Deere & Company established a presence in 1918 when it purchased the Waterloo Boy factory. Dubuque-built tractors were built at the company's factory in Dubuque, Iowa, which was established in the mid 1940s. The design of John Deere tractors differs according to where they were built (see chapter 1 for more on how and why). These differences in design translate into vast differences when it comes to service, restoration, and overhaul procedures, particularly as they relate to major components such as the steering, engine, transmission, and final drives.

In general, Waterloo-built tractors include the GP, D, A, B, H, G, R, 50, 60, 70, 80, 520, 620, 720, 820, 530, 630, 730, and 830. However, because of some differences in design, including the fact that the belt pulley ran off the camshaft, even the H requires a few variations in repair procedures.

The Dubuque-built tractors include the Models M, 40, 320, 330, 420, 430, and 435. For the purposes of this book, their predecessors—Models 62, L, and LA—are also considered Dubuque-built tractors, even though they actually originated in Moline, Illinois.

Like every company, John Deere was always looking for ways to improve its products. Often those improvements were implemented as soon as they were tested and proven, whether that occurred in the middle of a production year or midway through the production history of the model. For that reason, it is always best to refer to a quality service manual when making repairs, and to reference the applicable serial numbers for any repair procedure.

Models in the 30 Series, such as this 830 diesel, were built in Deere & Company's Waterloo, Iowa, factory. (Photograph by Hans Halberstadt)

Always Put Safety First

Certain safety measures are urged throughout this book. It's important to remember that tractors, wheels, and various other components are heavy. So be sure you take the proper precautions when lifting heavy items with a hoist or when blocking up the tractor.

Remember, too, that gasoline is highly flammable, as are paint mixtures, primers, solvents, and other chemicals you will be using during restoration. Do not have an open flame in the shop while working with any flammable substance.

Hopefully, you will find some shortcuts and time-saving tips in the following chapters. However, no amount of time savings is worth a compromise to your health and safety. Take the time to do it right and to do it safely. After all, tractor restoration is supposed to be enjoyable.

The LA was a forerunner of the Dubuque-built Models M, 40, 320, 330, 420, 430, and 435. (Photograph by Hans Halberstadt)

CHAPTER 1

Two-Cylinder History

Whether you refer to them as Johnny Pops, Johnny Poppers, Two-Lungers, or Poppin' Johnnies, John Deere two-cylinder tractors hold a unique spot in tractor history, not to mention the hearts of tractor enthusiasts world wide. Due in part to their unique and somewhat simple design, John Deere two-cylinder tractors are also a popular choice with tractor restorers and collectors. For some, it's that distinctive exhaust sound that only a two-cylinder engine can provide that draws their interest. For others, it is the memory of growing up with a two-cylinder tractor. They may not have driven it themselves, but they can certainly remember riding along with their granddad or uncle. Of course, there are a few who simply like the distinctive green and yellow color or who look at John Deere as the ultimate collector series.

Choosing the appropriate John Deere two-cylinder model to restore, however, can be a lot more difficult than simply deciding on the brand. During the thirty-six years that John Deere built two-cylinder tractors, the company introduced approximately twenty-eight different models, depending on which ones you count as unique models. Within those model lines are numerous sub-models and variations, including tracked versions, orchard models, and high-crop variations.

The Beginning of a Legacy

Prior to the introduction of the Model D, which was the first mass produced two-cylinder tractor with the John Deere name on it, the John Deere family tree actually splits into two different branches. Seeing the potential market for mechanized farming, Deere & Company began building prototypes for its own tractor models in the early 1900s. However, only one model ever went into production. Referred to today as the "Dain" tractor—named after Joseph Dain, the engineer who designed it—it went into production in 1918 with a run of somewhere between 50 and 100 units. With its four-cylinder engine and "reverse tricycle" design, it featured a single, chain-driven rear wheel that looked more like a drum than a wheel. Reportedly, only one Dain tractor has been located and restored. Today, it is owned by Frank Hansen of Rollingstone, Minnesota.

Meanwhile, the two-cylinder tractor entered the John Deere family by marriage. Like many early tractor manufacturers, Deere discovered that infrastructure was more important to early market establishment than tractor design. Market rival, Allis Chalmers, for example, purchased Rumely for the sole reason that Rumely already had a well established network of dealers. For the same reason, Deere & Company purchased the Waterloo Gasoline Engine Company the same year it introduced the Dain tractor—primarily to obtain a factory in which it could build tractors.

As part of the deal, though, Deere also acquired a commercially successful tractor—The Waterloo Boy—that had been in production for four years. Introduced in 1914, the Waterloo Boy had descended from a self-propelled tractor designed in 1892 by John Froelich. Although the Froelich tractor never became a commercial success, it did form the base for the

Referred to as the "Dain" tractor, this one-of-a-kind restoration represents the first tractor ever built by John Deere. This model, owned by Frank Hansen of Rollingstone, Minnesota, supposedly is the only one in existence.

Waterloo Gasoline Engine Company, which manufactured stationary, horizontal two-cylinder engines—even though Froelich himself had since left the company.

In 1911, with the hiring of two new engineers, the Waterloo Gasoline Engine Company took a renewed look at powered tractors in response to changes in the marketplace and developed the Waterloo Boy One Man Tractor. By the time the company was acquired by Deere & Company, the Waterloo Boy was already available in two models: the single-speed Model R and the two-speed Model N.

In the end, Deere decided to stick with the Waterloo Boy, instead of the Dain, for several reasons. For one thing, since the Waterloo Boy was less expensive to produce, Deere could market it at a lower price, which appealed to a greater number of farmers. With the purchase, Deere also acquired the outstanding orders for tractors, which meant immediate business. If John Deere executives needed a third reason to produce the Waterloo Boy, rather than the Dain, it was that the latter was never really a commercial success in the marketplace due to its weight, cost and cumbersome design.

Consequently, John Deere continued production of the Waterloo Boy until 1924. When Deere introduced the Model D in late 1923, calling it a 1924 model, the reign of the Waterloo Boy came to an end.

When the Waterloo Gasoline Engine Company was purchased by Deere & Company in 1918, the Waterloo Boy became the foundation on which John Deere built its tractor business.

The Poppin' Johnny Sound

One of the things about John Deere two-cylinder tractors that appeals to so many collectors is the unique and distinctive exhaust sound. This trademark popping started with the Waterloo Boy, which was the first two-cylinder, and continued through the production of the last John Deere two-cylinder tractor built in 1960.

The sound is due to the fact that all of the two-cylinder models, with the exception of the Model 435 (which used a two-cycle engine), utilized four-cycle engines. That means the engine rotates 720 degrees while going through a complete cycle. Why not have one cylinder fire every 360 degrees? As Gene Tencza, a John Deere enthusiast from Orange, Massachusetts, explains, "That would necessitate having the crankshaft throws for both pistons on the same side of the crankshaft at the same time. This means the weight of both pistons would always be shifting back and forth together, much the same as the weight of a single large piston."

One can just imagine how much that would cause the tractor to shake, says Tencza. Consequently, the engineers who designed the two-cylinder engine that was ultimately used in the Waterloo Boy came up with a different plan. During the course of the two revolutions required to complete the four-stroke cycle, the first cylinder fires at 0 degrees, the second fires at 180 degrees, then the engine coasts 540 degrees until the first cylinder fires again to begin the next cycle. In effect, it's like a person firing both barrels of a double-barreled shotgun in succession, reloading, and firing again.

It's easy to see the similarities between this Waterloo Boy engine and John Deere two-cylinder tractor engines that followed.

Thirty Years of D Models

As the first totally new tractor designed by John Deere, the Model D carried on much of the tradition of the Waterloo Boy, including the two-cylinder engine and a number of shared components. Like its predecessor, it was considered to be a wheatland tractor, designed for pulling plows and providing the power for threshers. (That was the roll of most farm tractors at that time, having picked up the jobs from steam tractors. For row-crop work, horses were still the primary mode of power.)

By collectors' standards, the Model D is still one of the most popular John Deere models. In addition to being the first completely new tractor offered by John Deere, it was the longest running production model in the company's stable. Over its thirty-year life span, from 1923 to 1953, when it was finally discontinued, some 160,000 units were built and sold. There were dozens of variations within the D line. Like many of the early John Deere models, the D was built in both unstyled and styled versions. In addition, there were numerous production changes in horsepower, features, and style that set them apart from each other.

The first fifty Model D tractors, for instance, had a 26-inch flywheel, as well as several other features that were not found on models that followed. The 26-inch flywheel and a one-piece steering shaft continued from serial number 30451 through 31279. Beginning with serial number 31280 and continuing through serial number 36248, a 24-inch spoke flywheel was used, along with a two-piece steering shaft. Finally, in 1925, with serial number 36249, a solid flywheel was adopted. Today, two-cylinder tractor enthusiasts still refer to early Model Ds as spokers, and those models are extremely collectable. Model D collectors can also tell you about the dozens of other variations that occurred over the years. They include differences in fender width, exhaust stack type, front axle configuration, and wheel type.

And therein lies part of the challenge and enjoyment of selecting a model to restore. Is the model you've located or have in mind just a run-of-the-mill unit or something truly rare?

Like the Waterloo Boy, the Model D was considered a wheatland tractor, designed for pulling plows and providing belt power to stationary equipment.

Although the Model D was built for thirty years, the early "spoker" models, like this one on display at Deere & Company's John Deere Pavilion, are the most sought after.

More Models Followed

The Model D may have been the first of a long line of two-cylinder John Deere tractors, but it wasn't the only choice for long. With the introduction of row-crop, general-purpose tractors by some of its toughest competitors, Deere couldn't ignore its need for a more versatile model. After two years of testing prototypes, the company released the GP in 1928. Initially introduced as the Model C, the GP featured a four-wheel configuration much like the D, but with an arched front axle and drop gear housings on the rear axle to give it row-crop clearance. It was also smaller, with a 10/20 horsepower rating compared to 15/27 on the D. Finally, it was the first tractor to feature a motor-driven power lift for raising its mounted cultivator and planter.

The GP had a few shortcomings, though, which prompted Deere to introduce the GPWT, or Wide Tread tricycle design, in 1929. The biggest feature of the redesign was the visibility, which was further enhanced in the 1932 version with an engine hood that tapered in the back. During its eight-year production run, various other versions of the GP were produced, including the GPO orchard version, the GP-P potato version, and crawler-tracked versions assembled by Lindeman Power Equipment Company in Yakima, Washington.

This model GP, restored during the winter of 2002, is the oldest model in a collection owned by Estel Theis of Savannah, Missouri.

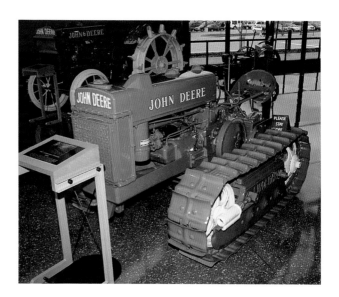

Through an agreement with Lindeman Power Equipment Company in Yakima, Washington, Deere introduced several tracked tractors, including this 1945 Lindeman BO (orchard). Deere later acquired the Lindeman company, which produced the tracked undercarriage, making it part of the Deere enterprise.

Deere Designs a Winner

Nothing sparked the sales of John Deere two-cylinder tractors like the introduction of Deere's second-generation row-crop tractors. This series began with the Model A in 1934. For once, John Deere had a model that could compete head-to-head with International Harvester's Farmall. Equipped with a tricycle, row-crop front end, the A also featured an adjustable rear-wheel tread width, a hydraulic power lift, and differential brakes that were geared directly to the large-diameter drive gears.

The A was rated for two 16-inch plows and produced 18.7 drawbar and 24.7 PTO/belt horsepower, thanks to its 5½x6½-inch distillate engine. Between 1934 and 1952, just over 328,000 units were built in various configurations, both unstyled and styled, although most were of the row-crop version.

There were still some people who thought the A was too large, so Deere responded with the Model B, which was largely designed to replace a team of two horses. Promoted as being two-thirds the size of the A, the B also weighed about 780 pounds less than the A and utilized two 4¼x 5¼-inch pistons to generate 11.8 drawbar/16 belt horsepower. Later versions featured more than double the horsepower. Because of its small size, low price, and versatility, the B seemed to be the perfect choice for a small farm. As a result, it became one of the best-selling tractors in John Deere's history. With production runs totaling 322,200 units, it was right up there with the A in popularity and sales.

In the 1930s, both the A and B were not only available in several specialized versions, but they were also the first Deere tractors available with pneumatic rubber tires.

The 1930s also saw a growth in the size of farms. As unfortunate producers were driven out of business by dust storms and the failing economy, the efficiency of mechanized farming allowed other producers to add acres and increase the size of their operations. In 1937, Deere responded to the needs of bigger farms with the three-plow Model G. Although it was similar in appearance to the A and B, it was substantially larger, heavier, and more powerful, with a horsepower rating of 20.7 drawbar and 31.4 PTO/belt.

This unstyled 1934 Model A on steel wheels was one of Deere's first successful row-crop tractor models.

This Model BR (regular) is just one of the many variations of the A and B models that Deere introduced between 1934 and 1952.

In the 1930s, the A and B were not only available in several specialized versions, but they were also the first Deere tractors available with rubber tires.

Like the BR, the AR was designed more for field work than row-crop applications. This AR was restored by Rex Miller of Avenue City, Missouri.

New York Styling Comes to the Midwest

One of the biggest changes affecting John Deere tractors, and one that established the greatest demarcation for collectors, occurred in 1938, when Deere introduced the newly streamlined and styled Model A and Model B. The transformation actually began in 1937, when Deere hired Henry Dreyfuss Associates, a Madison Avenue design firm well known in the industrial design world, to help style its two most popular models.

Born in New York, Henry Dreyfuss had apprenticed as a designer of theater costumes, scenery, and sets, before establishing himself as an industrial designer. Tractors were about as foreign to the New York designer as Madison Avenue was to most farmers. That didn't deter Dreyfuss, however, from doing what he called a "clean up" on the John Deere tractor design. In his initial efforts, he produced a redesigned hood that surrounded the steering gear and developed a grille that enclosed the radiator.

Although the new styling was eventually extended to the M and the D, the first Deere tractor to feature the new styling from its inception was the Model H, introduced in 1939. Billed as a two-row tricycle model, it featured 3½x5-inch pistons and produced 12.5 drawbar/14.8 belt horsepower. Built from 1939 to 1947, it was produced in numbers exceeding 60,000. The H had a few peculiarities. One was that power was taken off the camshaft, rather than the crankshaft. This meant that the belt pulley ran in the opposite direction from that of crankshaft-driven belt pulleys. It also meant that the bull gears were not required, which in turn required the brakes to be applied directly to the axles.

With the exception of their amenities—like electric starters, lights, Powr-Trol hydraulic control of pulled implements, and padded seats—horizontal two-cylinder tractor models remained pretty much unchanged through much of the 1940s. The one exception came in 1949, when Deere introduced the Model R, the company's first diesel tractor. The most powerful tractor to date, the R featured a 5¾x8-inch diesel engine that generated 45.7 drawbar and 51 PTO/belt horsepower.

There's quite a contrast in size between the Model H (left) and the Model G (right). The 15-horsepower (belt) H was Deere's smallest Waterloo-built model, while the three-plow G was the largest for its time. Both carry the styled hood and grille developed by Henry Dreyfuss in 1938.

A Deere of a Different Kind

In 1937, John Deere introduced a tractor that was dramatically different than the other models in the line. Billed as a one-plow, utility tractor, the Model 62 tested at 7.01 drawbar and 9.27 PTO/belt horsepower. More unusual than its low horsepower rating, though, was its vertical two-cylinder engine. Unlike the other John Deere tractors of the time, it had a driveshaft and a foot clutch. Moreover, it wasn't built in Waterloo, like the other models, but at the company's Moline Wagon Works factory in Moline, Illinois.

Of particular interest to collectors, however, is the low production run of the Model 62. Only seventy-eight units were built before it was redesigned, bumped up by four horsepower, and released as the Model L. A few years later, in 1941, John Deere released the slightly larger Model LA. Not only did it have a slightly larger engine, but it also featured a 540 rpm PTO in addition to a belt pulley. Other than the PTO and larger rear tires, though, it is pretty hard to distinguish the LA from the L. Both models were discontinued in 1946.

In 1947, Deere introduced the two-cylinder, 18.2 drawbar and 20.5 PTO/belt horsepower Model M as a replacement for the L and LA, as well as the H. At this point, the company also opened a new plant in Dubuque, Iowa, which became the manufacturing point for all succeeding vertical two-cylinder tractors. Two years later, the company added the MT model. In terms of horsepower and features, it was almost the same tractor as the M, except the MT could be equipped with adjustable front axle, dual tricycle front, or single front wheel.

Introduced in 1937, the Model L was one of the first vertical two-cylinder tractors built at the John Deere Wagon Works in Moline, Illinois. It was designed to be an inexpensive replacement for a horse team.

This industrial version L, built in 1940, is just one of the many tractors that was restored by the John Deere Collectors Center for display in the John Deere Pavilion.

Rated at 21 PTO horsepower, this Model M, owned and restored by Rex Miller of Avenue City, Missouri, still makes a good tractor for pushing snow off the driveway.

With the switch to the two-number designation in 1953, the Dubuque-built Model M became the Model 40. Along with the new number came numerous improvements and a 15 percent boost in horsepower.

Numbers Replaced Letters

After World War II, Deere addressed the growing tractor market in earnest, introducing six new models in ten years time. Having introduced the R as an eventual replacement for the D, the company moved on to produce a new numbered series in 1952. It began with the Models 50 and 60 as replacements for the B and A respectively. In addition to new sheet metal styling that featured small vertical grooves in the radiator screens, the numbered series introduced duplex carburetion for more power from each cylinder. A year later, Deere introduced the Model 40. It had the vertical cylinder engine that powered the M but got about 15 percent more horsepower out of it. Although it was built in the Dubuque factory, the Model 40 actually shared its styling and features with the tractors built in Waterloo. In 1954, Deere replaced the G with the Model 70, which was available with a gasoline, "all-fuel," LPG, or diesel engine; the latter option made it the first diesel row-crop tractor to carry the John Deere name. Finally, in 1955, Deere replaced the R with the diesel Model 80.

In 1956, the company picked up even more speed and power by introducing the 20 Series. Built in the Waterloo plant, the tractors featured a two-tone paint scheme that was highlighted by a yellow stripe that ran along the base of the hood and continued down the sides of the radiator shroud. The 20 Series also introduced the new "Float-Ride" seat, which became standard equipment on all units. Available in six models—the 320, 420, 520, 620, 720, and 820—the 20 Series featured new engines that offered increased combustion and performance from newly designed cylinder heads and pistons.

Deere only kept the 20 Series in production two years. In 1958, all six models were replaced by a comparable 30 Series model. Although very little was changed internally, ease of use and operator comfort took on new meaning. One of the highlights was a new dash, which featured easier-to-read instruments and a steering wheel column positioned at an angle. The tractors also featured new flat-top fenders with built-in dual lights.

One year later, Deere introduced the Model 435, a diesel tractor powered by a General Motors two-

The Model 50 was one of several models introduced in 1952 as replacements for the letter series tractors. In this case, the 50 replaced the styled B.

Introduced as a replacement for the Model G, the Model 70 came out in 1953—a full year after the Models 50 and 60.

cylinder, two-cycle engine rated at 28.4 drawbar horsepower and 32.9 horsepower at the PTO and belt pulley. The Model 435 was extremely short-lived, however, having a production run of only one year and 4,488 units.

On August 30, 1960, John Deere surprised the industry by unveiling the "New Generation of Power" tractors. The models included the well-respected 1010, 2010, 3010, and 4010. Unlike John Deere tractors built up to that date, the tractors were powered by all-new vertical four- and six-cylinder engines. Gone was the two-cylinder engine and the familiar popping exhaust resonance that only a John Deere two-cylinder tractor could provide. It was the end of an era and a time that many two-cylinder tractor restorers and collectors still remember as a high point in John Deere history.

The Model 420 joined the other 20 Series models in 1956 in eight different configurations, including row-crop, high-crop, crawler, standard, utility, two-row utility, and two specials. Similar to the 320, it boasted a larger engine with 113.5 cubic inches and 23.5 PTO horsepower.

Like the other 20 Series models, the 520 offered more comfort features than its predecessor, including a new Float-Ride seat. Another important feature was the Custom Powr-Trol three-point hitch, which offered draft control and the ability to preset the working depth.

Above: Just as it did with earlier models, Deere continued to offer the 20 Series in a variety of configurations, including this "regular" owned by Harold Hatfield, of Salsbury, Missouri.

Left: Although the Model 620 came in a number of configurations, the conventional row-crop model, like this one owned by Estel Theis, represented more than 80 percent of the production.

Also built in Dubuque, the 430 replaced the 420 in 1958. Like the other 30 Series models, it featured a slanted dash and steering wheel, along with other operator comfort features.

Above: By 1958, when this Model 730 was built, diesel engines were becoming a popular choice among tractor buyers. This 730 is owned by Everett Ragsdale, of New Hartford, Iowa.

Left: Built for less than two years, the Model 435, was not only the last two-cylinder model to be introduced by John Deere, but it was also the only model with a two-cylinder, two-cycle diesel engine—making it both rare and collectible.

CHAPTER 2

Shopping for a Tractor

For some tractor restorers, shopping for a tractor is the adventurous part of the project. Where you start your search, though, depends a lot on your goals.

If you're buying a tractor strictly as a collector model, a whole different set of rules tends to apply. The rules depend, too, on whether you're collecting the model for its value on the market or its sentimental value to you alone. As an example, some tractor restorers simply have a desire to restore and collect all the models within a certain series, such as the letter series models or the 20 Series. Perhaps you just want to own a model like the one you drove while growing up on a farm or visiting a grandfather. If that is the case, it's better to evaluate the tractor as if you were purchasing a model for doing work around the yard or farm. Since many of these models are not rare, it's best to look for a combination of sound mechanics, good cosmetics, and a reasonable price.

On the other hand, if you intend to buy a vintage tractor as an investment, you should plan on doing some research and perhaps checking out the credibility of the seller and model being represented. While most enthusiasts are honest people who love John Deere tractors as much as you do, there are people who will intentionally or unintentionally misrepresent the products they have for sale. If you're interested in a particular model, study factory literature or tractor books to find out how many of that particular model were built. This will give you an idea how rare that model is and what it might be worth when you are finished. Find out, too, if there are any distinguishing characteristics of the tractor that might identify it as being the real thing—even if sheet metal or certain components have been changed.

Chris Pratt, with *Yesterday's Tractors* on-line magazine, explains that working with rarities almost always rules out looking for perfect mechanical and cosmetic condition. "I have see extremely rare tractors purchased that consisted of just the engine block, rear end, rims, and frame assembly," he says. On the other hand, Pratt cautions against buying a tractor on which the cosmetic components are the only thing that makes that specific machine rare.

"A common example of this is some orchard model tractors," he explains. "Frequently, there are no remnants of the orchard add-ons or anything but a model designation to distinguish the machine from its common utility version brother. Finding orchard models may be relatively easy, while finding the orchard components that make your project collectible is next to impossible. If an incomplete model is priced as a rarity, it may be wise to pass."

If you're shopping for a collector tractor, you should also take a look at the tractor's serial number. As a general rule, the lower the number, the greater the tractor's collector value. A high serial number might indicate that the tractor was one of the last models of its type to come off the assembly line. One example is the last ninety-two Model D tractors to be built. Known as "streeters," they were literally built in the street or roadway after the production line had already been disassembled and changed over to another model.

Again, it helps to know some history of the serial numbers assigned to the model you're inspecting. Normally, the first ten and the last ten units within any particular model run have more value than the rest of the units within that model run.

Should you be fortunate enough to find something like a 1958 630 LP Standard, however, it really doesn't matter what serial number the tractor carries. Since there were only sixteen built, any one of them is going to be very collectible.

With literally hundreds of tractors to choose from, the selection of a vintage tractor for restoration comes down to personal preference and your plans for the finished product. This line-up, for sale in central Missouri, provides three choices.

Some antique tractor shows have begun featuring a vintage tractor auction, which is one source of a restoration project.

The goal of many restorers is to simply add to their collection, which often encompasses one favorite brand. This assembly of 20 and 30 Series models is found at the John Deere Collectors Center in Moline, Illinois.

One of only ninety-two units built, this 1957 620 LP Orchard tractor is a true collector's model. You need to be careful about paying too much for something like an orchard model, though, if the sheet metal that makes it unique is badly damaged or missing.

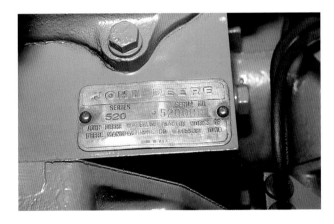

The serial number plate alone can add value to a tractor, especially if it proves that the model was one of the first or last in the series to be produced.

Search the Records

One of the nice things about John Deere tractors is there is still a wealth of information about them available from both the company and collector clubs. As an example, the Two Cylinder Club, based in Grundy Center, Iowa, has already published production registers for the unstyled B, unstyled G, and Waterloo-built 30 Series tractor. Other registers are planned for the near future. Each register contains production information on thousands of tractors within the respective line, including serial numbers of each model and variations that were built.

In addition, the Two Cylinder Club will provide serial number research on the majority of two-cylinder tractors built in the Dubuque and Waterloo factories. At the time of this writing, the fee for one search was ten dollars. However, since it takes less time and effort to do multiple researches, the club offers a discounted rate when more than one research is requested on any single order. The production information available varies considerably from one model to the next, and often from tractor to tractor within a given model. When possible, the club will provide the build date—which may not actually be the date the tractor was built, but rather the date it was entered into the inventory records—the ship date, the specific shipping destination, and the optional equipment installed on that model.

Tractor build records are also available from Deere & Company for nearly all John Deere two-cylinder tractors built since 1914. To access the Deere records, contact the John Deere Collectors Center (listed in the appendix) with the serial number and model of the tractor.

Some collectors have used the build records in reverse. Instead of requesting information on a tractor they have purchased or are looking at, they've used build records to seek out rare models. There's a story, for example, of one collector who was searching for a hard-to-find Waterloo Boy tractor. After researching the records, he traced one to a branch house in Montana. After tracing it to the general location it was sold, his search moved from paper to shoe leather. One day, while stopping for gas in a small country service station, he casually mentioned the search to a few older gentlemen playing cards. One of them responded, "I think I know where that tractor is." With the local historian along as a guide, the would-be restorer found the very Waterloo Boy tractor he was seeking. It was buried in a tree row, totally out of sight of passers-by on the nearest road.

Similarly, a number of enthusiasts in search of a high-crop model start with the records, which often lead them to Louisiana or another area known for sugar cane or high-clearance crops. Once they locate the general area where the tractor was sold, they begin looking up people who may have worked at the local John Deere dealership, including retired shop mechanics. Sooner or later, someone seems to know where a particular tractor is or what happened to it.

Any high-crop model is highly collectible. In fact, some collectors spend hours going through records and searching the back roads to find models such as these high-crop Gs.

How to Tell Them Apart

To the inexperienced tractor enthusiast, many of the John Deere two-cylinder tractors look alike or very similar, especially when the only decals that identify a model as an A or a B have long since worn away. Unless you see one model setting next to another model that is larger or smaller, there's really nothing to compare the size to for identification.

It is fairly easy to tell the difference between a styled and an unstyled model. The latter essentially has no sheet metal parts except for a simple hood that covers the fuel tank. Styled tractors feature a grille in front of the radiator and a hood that not only covers the fuel tank and part of the steering rod, but also blends into the grille. But how do you properly identify an unstyled tractor setting in a farmyard or an old fencerow?

According to Gene Tencza, a John Deere enthusiast from Orange, Massachusetts, the easiest method—when checking the serial number plate is not possible—is to focus on the intake stack, the exhaust stack, and the steering post.

The Model G, he explains, has two stacks located side by side, and the steering post on the unstyled version has a removable plate on the front, attached with four bolts.

The Model A also has the stacks located side by side, but the steering post is smooth in the front and has the bolts in the back.

Finally, on the Model B, the stacks are side by side, but the exhaust stack is located slightly ahead of the intake stack. The steering post has the bolts in the front, but the cover plate only has two bolts.

"This method is foolproof, even if the stacks are missing and you can't get close enough to count the bolts," says Tencza. "You should still be able to see the holes in the hood were the stacks should be. The exhaust stack is always on the right if you are sitting in the tractor seat."

While there's no decal to confirm your suspicions, the two bolts on the steering column cover plate confirm that this is an unstyled Model B.

True John Deere collectors know that the front axle shaft on early Model A and B tractors was mounted with four bolts. Later models used eight bolts for greater strength.

Buying the Right Tractor

The time you spend looking for the right model and inspecting each tractor will pay off later in the form of greater efficiency, less time and money spent on restoration, and increased satisfaction with the finished product. If it's a work tractor you're looking for, the benefits may also include increased safety; buying a tractor that is not ideal for your needs may be not only inefficient but also dangerous.

Another aspect to consider when looking for a prospective project is the geographic location where the tractor was used. You'll quickly find that, as a general rule, tractors used in the eastern and southern parts of the United States are more apt to have rust problems or a stuck engine, while those from the western part of the country tend to suffer more tire damage due to sun exposure and dry rot. Humidity has a tendency to take its toll, and nothing is worse than salt air.

Estel Theis, a John Deere enthusiast from Savannah, Missouri, agrees that the dryer climate prevalent in the continent's High Plains helps preserve the classics. Nearly half of the vintage models he has restored to date have come from farms in Montana. He insists that, in addition to having less rust, western tractors exhibit less wear on the steering mechanism and front axle, simply because the fields are bigger and flatter and the tractor has made fewer turns.

Theis agrees that it's helpful to know the history of the tractor you're buying. If, for example, the tractor was used to haul or load manure in a livestock operation, you can expect to spend some money on front wheel bearings and seals. Likewise, the steering mechanism on a row-crop tractor is likely to require more work than that of a wheatland version.

There are other things that can tip you off to potential problems. One is the hose between the carburetor and the air cleaner tube: If it is cracked or missing, the engine may have sucked in a lot of dirt and will need work. Likewise, a missing or cracked shift lever boot can let water into the transmission—and nobody has to tell you what kind of problems water can cause when it freezes and thaws repeatedly.

In contrast, antifreeze in the cooling system and oil in the oil-bath air cleaner tells you right away that the tractor was treated well and is likely in good shape.

If the tractor is in running condition, there are several things you should check both before and after you start the engine. In fact, you may want to turn to chapter 5, "Troubleshooting," and run through some of those procedures before you make your final decision.

If the tractor is not in running condition, you'll want to make sure the pistons are not rusted to the cylinder walls. One way to do that is to pull one or both of the spark plugs and shine a light inside the cylinder walls to check their condition. If the walls are shiny, the pistons are probably not stuck too badly, if at all. On the other hand, if you can't even get the spark plugs out due to rust, you might have cause for concern.

Negotiating a Price

Before you start dealing on any tractor, you should know your needs, your budget, and what is on the market. Become as knowledgeable about the prospective tractor as you can through research, conversation with other collectors, and physically checking it out.

This research will allow you to go into the negotiations with a price in mind. If your preview of the tractor turned up any problems, you may find that the seller is willing to come down on the price. However, you need to decide if you have the time and the expertise to correct what you have found.

You may find that the tractor won't start or run on the day you look at it. The seller may, in all honesty, tell you that everything worked fine when he or she last drove it; but when a tractor sits for very long, it can develop problems that even an owner is not aware of. In this case, you must start your bidding from nearly scrap level prices, since you have no idea what you're getting into, then go only as far as your conscience and experience will allow.

Many collectors also go into a negotiation with the idea that a stuck engine is basically junk. If it can

Some restoration projects are no more than a pile of parts when purchased. Obviously, the price you're willing to pay for such projects should depend on the rarity of the model.

be freed, that's a bonus. Hence, your offering price should reflect that possibility.

Finally, know how much you are willing to spend on the whole project before you start negotiations. Many people don't realize how much expense is involved in a restoration. Even if you get the tractor at a decent price, you have to anticipate the cost of engine repairs, body work, new wiring, replacement components, and so on. Many restorers suggest that you assume the worst of any potential restoration. That way you won't be surprised if you can't find parts or if it takes more time and money than you expected to get the job finished.

Above: If a tractor is not in running condition, as is the case with this A and B, you almost have to start your bidding price based on its junk value.

Right: When shopping for a tractor, it helps to know what was original equipment. Obviously, the decals and seat have been changed during the life of this steel-wheeled model. The skeleton wheels, however, make it collectible.

It's unusual to find a model that has dual wheels on the back and only a single front tire. This particular model is owned by Bill Batt from Garden City, Missouri.

Wheatland models, such as this 730 diesel, are far more rare than row-crop or standard wide-front models.

Special features, like these fender extensions on a Model R, can add a tremendous amount of value to a tractor.

CHAPTER 3

Setting Up Shop

Basic Tools

If you're serious about doing a first-class tractor restoration, the first thing you're going to need is a good set of tools. And make sure you put the emphasis on *good*. Cheap tools are only going to lead to frustration as they break, strip, or, worse, damage a tractor part. Look for a set of automotive-quality tools that come with a warranty, such as those offered by Sears (Craftsman), NAPA, and Snap-On.

You'll want to start with a drive socket set that contains sockets ranging from /° inch to over 1 inch. In addition to a ratchet handle, you'll need a breaker bar to loosen stubborn bolts without risking damage to the ratchet.

A set of combination wrenches will come in handy, too. There are some places you simply can't get a socket and ratchet into. You can decide which will work best for you and your budget, but choices include open-end, box-end, and wrenches that provide an open-end configuration on one end and a box-end of the same size on the other.

To round out your tool collection, you'll want to add a couple of adjustable wrenches (often referred to as Crescent wrenches, even though Crescent is a brand name), a full set of regular and Phillips screwdrivers, a pair of adjustable pliers, needle-nose pliers, and a pair of locking pliers (often referred to by the popular brand name Vise-Grip). Other tools that you'll probably need at some time or another include a good hacksaw, a punch set, and a cold chisel. And don't forget to pick up a couple of putty knives. You'll need those for scraping away grease and grime.

Don't assume you have to go out and buy all new tools. Due to the economy and changes in agriculture, farm auctions are far too common these days. So keep your eyes open for an estate sale or a shop liquidation where you can pick up what you need at a reduced price.

Above: While your work area doesn't have to be anything fancy, it does help to have a place where you can keep the tractor indoors while it is being restored, especially when it comes time to paint the tractor.

Left: A good set of tools is almost a prerequisite for a quality vintage tractor restoration.

Specialized Tools

Depending upon how much engine or electrical work you get into, there are other tools that you may need. These include feeler gauges, a point dwell meter/tachometer, voltmeter, and compression and vacuum gauges. For tools and gauges that are only used on occasion, consider renting or borrowing from a fellow restorer.

Other tools that can make your life a lot easier include a gasket scraper, bearing and hub pullers, and a seal puller. A pickle fork can be a handy item, too, if you plan to separate the tie rods on the front axle.

If you anticipate doing much engine work, you'll also need tools for making precise measurements. At a minimum, these should include a micrometer for measuring items up to approximately an inch in width; a dial caliper to determine the acceptability of parts such as the crankshaft and camshaft; a set of feeler gauges; and a dial gauge for measuring certain types of end play.

You'll also need a torque wrench. While most parts of the tractor don't require a specified torque rating, you'll find that many engine components, including the head bolts, must be tightened to a particular setting, in order to reduce the chance of head warping and oil and water leaks. While there are several different types of torque wrenches, perhaps the easiest to use is the type that makes an audible click when the correct torque rating has been reached. Since you preset the desired torque with a dial on the wrench, you don't have to worry about providing enough light or room to read a scale or being able to see the scale.

Of course, you can get by with the older style of torque wrench that uses a stationary pointer and a gauge attached to the handle. As the bolt or nut is tightened, the handle bends in response to the applied torque. As a result, the needle, which is fixed to the socket head, moves up or down on the scale to indicate the amount of torque being applied. You just have to be able to watch the scale as you're tightening the fastener.

Depending upon how far you get into engine repair, you may also need specialized tools such as a ridge reamer, valve-lapping tool, piston ring compressor, and cylinder hone. The use of each of these items is discussed in more detail in chapter 6, "Engine Repair and Rebuilding."

A number of specialty tools, including such things as feeler gauges and valve lapping tools, can be found at a reasonable price at your local automotive parts store.

A torque wrench is just one of those specialty tools you'll need to purchase or borrow when overhauling an engine.

An oxy-acetylene torch can be valuable for cutting metal and heating stubborn parts that refuse to budge.

If you plan to do your own painting, an air compressor with adequate capacity, an in-line water filter, and plenty of hose will make the job a lot easier. You'll find compressed air equally helpful when cleaning parts and components.

Air Compressor

While an air-powered impact wrench can be a valuable asset when removing stubborn bolts, air-powered tools aren't quite as necessary as the air compressor itself. You'll want a portable air compressor with a tank for a couple of reasons. First of all, an air hose and nozzle are invaluable for blowing dust and dirt out of crevices and away from parts. You'll want to use it to blow out fuel lines, water passages, and the like, as well.

Secondly, assuming you're going to be painting the tractor yourself, you'll need an air compressor to operate the paint sprayer. Here, tank capacity is important. If the tank doesn't have enough capacity and the pump can't keep up, you're going to be painting a few minutes, stopping to let the pressure build in the tank, painting a few more minutes, and then waiting again. Hence, you might want to put things in reverse order and shop for a paint sprayer before you look for a compressor.

Most restorers who do their own painting suggest using a compressor with at least ½ horsepower that is capable of delivering at least 4 cubic feet of air per minute at 30 psi pressure.

Vise

You don't have to own a vise to do a tractor restoration, but considering the availability and reasonable cost, and the versatility that a vise provides, you'll likely find it worthwhile. Just being able to clamp a part in the vise while you work on it can be helpful at times. And you'll be especially glad you have one when you've got a part in which a bolt absolutely won't budge.

For the most versatility, many shop owners recommend at least a 6-inch vise that is bolted securely to a solid bench. You may even want to get a piece of plate steel to attach to the bottom side of the bench for extra strength and support.

Anvil

Another tool that you'll at times find invaluable—especially if you have to straighten sheet metal—is an anvil. You don't have to invest in a commercial shop anvil, though. For what you'll need most of the time, a 2- or 2½-foot piece of railroad track rail will do. In fact, the rounded edge of the rail will work better than a real anvil for some metal fabrication.

If you occasionally need a flat surface or a square end for bending, you can weld a piece of bar stock across one or both ends of the rail so it will stand upright when turned rail side down.

Hoists and Jacks

Last, but certainly not least, you're going to need equipment to lift and support the tractor, engine, and other components. Naturally, your needs will depend to some extent on the type of tractor you're restoring. Some models, such as the M and MT, use the engine as a load-bearing member; in other words, there is no full-length frame. The front of the tractor attaches to the front of the engine, and the rear of the engine attaches to the transmission. To remove the engine for an overhaul, you have to split the tractor in half. That means you'll need a floor jack or bottle jack to support each half of the tractor.

Most Waterloo-built tractors, on the other hand, employ frame rails that attach to the transmission. The engine then sits within this frame, which means that the entire engine can be lifted out while the tractor remains on its wheels. This type of arrangement, however, generally requires an overhead hoist or an engine hoist to lift the engine out of the frame. One exception is the Model D, which has the front axle support attached to the cylinder block. It will be necessary to support the front of the tractor under the main case until the support is detached from the block.

Unlike most other Waterloo-built tractors, the Model D utilizes the engine block as a support for the front axle. Hence, its axle is not quite as easy to remove for overhaul as that of an A or B.

A hoist of some kind is a necessity when removing the engine head or splitting a Dubuque-built tractor for engine overhaul or transmission work.

Even if it is just done on a temporary basis, placing the tractor on cribbing, or wooden blocks, will make it easier to work on the axle and steering components.

There will come a time, too, when you need to remove the rear wheels, the front axle, and the front wheels for cleaning, restoration, and painting. Again, you'll need some heavy-duty lifting equipment capable of raising the tractor to the level where you can block it up on stands or wooden blocks. Don't try to get by with concrete blocks! They can crumble or crack too easily, posing a physical danger. Don't try to pile blocks up too high, either. The best bet is to build cribbing under the frame, which means you place strong wooden blocks log-cabin style under the tractor or axles as structural support until the wheels can be safely reinstalled.

Some restorers like to block up the entire tractor from the start, pull the wheels, strip the tractor down to the frame, and work on restoration from the ground up. Others like to leave the tractor on its wheels as long as possible, work on components as they go, and roll the tractor out of the way when necessary.

The bottom line is your lifting and cribbing needs will depend largely on the size and type of tractor you're restoring and how you prefer to work. Just be sure that you keep safety in mind and that you have the right equipment to do the job. Don't try to lift the rear end of the tractor, for example, with a single bottle jack, even if it is an 8-ton jack. That's not what you would call adequate stability.

Purchase a Good Shop Manual

Considering the number of different two-cylinder John Deere tractor models built between 1924 and 1960, and the number of sub-models within each model, this book cannot go into detail on each and every model—especially with powertrain restoration. There is simply no way to cover all the specifics and idiosyncrasies. You'll also need a source of specifications such as tolerance limits, torque settings, and wear limits. Therefore, you need to purchase a repair manual for your specific model.

There are a number of good sources for service and repair manuals listed in the appendix. Intertec Publishing, for example, offers a complete line of their I & T (Implement & Tractor) Shop Service manuals, which are available for virtually all John Deere tractors you're likely to encounter.

You might check with your local John Deere dealer as well. Depending upon the age of the tractor, an original manual may still be available.

Of course, if you want to pay the price, you can still find original service manuals for older tractors for sale by vendors at a number of flea markets, tractor shows, and swap meets.

Before you start a restoration, one of the first things you'll need to obtain is a good service manual for your tractor model. If for nothing else, you'll need it to determine tolerance limits and specifications.

CHAPTER 4

Getting Started

Take Your Time

The first step in restoring a vintage John Deere tractor, once you have it home, is to convince yourself that it's going to take some time. A lot of the people restoring antique tractors are retired farmers or mechanics who do it for the enjoyment of seeing an old tractor brought back to life. Some of the other restorers featured in this book are full-time farmers who spend much of their winters working on a tractor. However, once spring arrives, their pet project tends to sit until ground preparation and planting are complete.

The point is, unless you're retired or have three or four cold winter months to devote to the project, a quality restoration is going to take up to a year or more. Trying to finish the project too quickly is either going to lead to discouragement or dissatisfaction later on with the shortcuts you have taken.

Establish Your Goals

The second thing you need to realize is that *tractor restoration* can mean lots of different things. Hence, one of the things you'll need to do right up front is decide how far you want to take the restoration project. To some enthusiasts, a vintage tractor restoration is nothing short of restoring the tractor to mint condition. That means they go through the engine, transmission, rear end, and every other component that might need attention. They also insist on accuracy in every detail.

On the other hand, not everyone has the budget to do a first-class restoration. If you're in that category, you need to decide along the way what you can live with and what you can't. As an example, if you only plan to drive your finished product in a few parades a year and take it to a few antique tractor gatherings, you may not need to overhaul the transmission. But if you plan on using the tractor to mow the roadsides, plow the garden, and push snow in the winter, you'll need to make sure both the transmission and the final drives are in good working order. By the same token, you'll have to decide whether you want to use substitute components like modern tires for the original style, screw-type hose clamps for wire clamps, and black-faced gauges for white-faced reproduction gauges. All contribute to the price of your project while increasing the authenticity. The question is, how authentic do you want the tractor to be?

Don't try to cut costs where it doesn't make sense, though. If there is one common lament among tractor restorers who make their living restoring tractors for paying customers, it's that some clients don't want to spend the money to do it right the first time.

While it would be an ideal work tractor, this wide-front John Deere 720, lovingly restored by Dennis Funk, from Hillsboro, Kansas, is meant only for show.

Clean It Up

Unless the previous owner was meticulous about the care of the tractor, you can't expect to find a tractor that isn't covered with a thick layer of grease, dirt, and grime, particularly after fifty to seventy years of use and abuse. That's assuming it even runs and has been protected from the elements for a good portion of its life. If you're restoring a tractor that has been sitting in the fencerow for the last twenty years, you're probably looking at a lot of rust, too.

Consequently, the first step in any tractor restoration is trying to find the potential that is hidden beneath years of neglect. After you pick up the tractor, you can start by running it through a car wash on the way home, unless you have a hot pressure washer of your own. Hot water or steam cleaning is going to do the best job of removing oil and grime.

There are several methods that you can use for removing grease, paint, and rust, and each has its own place, along with advantages and disadvantages. Bill Anderson, a full-time tractor restorer from Superior, Nebraska, likes to spray oven cleaner on some of the really greasy areas and let it soak before power washing the whole tractor. Others use a hand-pump sprayer to soak the tractor and engine with diesel fuel for several hours. Spray-on engine degreaser can be an asset, as well.

Chances are, though, it's going to take more than hot water and engine cleaner. Grease that has mixed with dirt and debris, and become baked on by engine or transmission heat, can get as hard as a rock. So you'll want to keep a putty knife and wire brush handy while you're cleaning. You'll find some of the low-cost "tools" lying around the house to be equally handy. Old toothbrushes are especially useful for scrubbing delicate parts; cotton swabs can be dipped in paint thinner to clean hard-to-reach crevices; and pipe cleaners can be used to clean the channels and tubing.

If you're doing a complete restoration, you should start disassembling the tractor at this point anyway. Start by removing the major components, like the hood, fenders, fuel tank, and so on. Not only will it make the job of mechanical restoration easier, but you will be able to do a more thorough job of cleaning and painting the parts if they are already off the tractor.

While cleaning the tractor and stripping paint, keep in mind that any traces of grease, oil, rust, and old paint can create problems with paint adhesion. The goal is to make sure every surface will hold the paint you will be applying later.

It is not recommended that you use gasoline or kerosene as a cleaner. There are cleaners and degreasers available at any automotive store that are both safer and more effective.

The first step in tractor restoration consists of stripping the tractor down and cleaning it up. Obviously, this Model G is going to take some work.

Your local automotive parts dealer should be able to direct you to a wide assortment of paint strippers for cleaning sheet metal down to bare metal. Many restorers say they have the best luck with Aircraft Remover (far right).

In the early stages of tractor clean-up, a simple putty knife can be a valuable tool for removing baked-on grease.

While it's not as easy as using a sandblaster, Estel Theis of Savannah, Missouri, finds that a wire brush on an electric grinder can take off years of accumulated rust.

Disassembly

Either before, during, or after cleaning, you'll need to start tearing the tractor apart, beginning with things such as the fenders, fuel tank, grille, and hood. Remember, it may be a year or more before you're ready to reassemble the tractor. Consequently, it's important to maintain good records as you disassemble the tractor. Keep a pad and pencil handy for recording measurements and taking notes. Take your time, labeling parts, if necessary, so you can remember how they fit back together. Used in combination with a service manual, the notes and images you record today will be a valuable resource several months from now.

You'll want some good photos, even if you don't need them for reference later on. One of the first questions people are going to ask you is, "What did it look like when you started?" Showing them photos of the "iron pile" you dragged into the shop is half the enjoyment.

Another thing to consider is the number of bolts, nuts, and washers you're going to need to keep track of. One way to organize them is to collect a bunch of egg cartons and put the nuts and bolts from different areas into individual egg compartments. You can even use and label separate egg cartons for different parts of the tractor (one for the grille and hood, one for the transmission cover, and so on). For bigger bolts or parts, you can use larger containers such as coffee cans and plastic butter tubs.

Whatever you do, don't throw anything away. Even if a piece is rusted beyond any possible use, it may be needed as a pattern for creating a new piece later on.

Finally, be careful about using too much force when trying to remove rusted or frozen parts. In your haste to break things loose, it's easy to damage irreplaceable parts. Quite often, the best bet is to use a combination of penetrating oil, patience, and a properly sized tool.

If the part can withstand the heat, a propane or oxy-acetylene torch and occasional taps with a hammer can be as effective as anything. Alternately using heat and penetrating oil can also be helpful. Just don't apply oil to hot metal or direct an open flame toward a pool of penetrating oil. Be careful, too, about using

Although this John Deere two-cylinder tractor has extensive fire damage, the sheet metal appears to be in good condition. Since you can still see where the decals were located, it would be a good idea to take measurements before stripping and cleaning the panels. These measurements will make the job a lot easier when it comes time to position new decals.

a torch on a part where the heat can be transferred to a bearing. Using a torch to loosen the flywheel on a horizontal two-cylinder tractor is a good example. Unless you know you're going to be replacing the driveshaft bearing, the time you save may not be worth the cost.

Finding Replacement Parts

Before you get too deep into the restoration project, you'll need to consider the challenge of locating and acquiring replacement parts. As you're disassembling the tractor for cleaning, begin making a list of all the parts you'll need to restore the machine to show or working condition. Record as much detail as possible, including part dimensions, their shapes, and any serial numbers listed on separate parts. This list will give you an idea of how much you may need to spend on bearings, gaskets, sheet metal components, and so on. It will also give you a head start on locating some of those parts. By knowing what you need ahead of time, you can be searching the swap meets, salvage yards, and classified ads for the necessary components while you're working on other areas.

The good news is that antique tractor parts are much easier to find today than they were in the past, thanks in part to the growing interest in tractor restoration. In addition to the new parts that are still available through John Deere dealers, there has been a proliferation of small companies that specialize in restoration parts. Through the sources listed in the appendix and in many of the restoration and tractor club magazines, you can find everything from reproduction grille medallions and rubber torsion springs to fenders and flywheels. There are numerous individuals and companies, too, that can repair your old magneto, carburetor, fuel injector pump, or distributor.

"John Deere tractor restoration is far easier today than it was just ten or twenty years ago," says Jeff McManus, senior consultant for the John Deere Collectors Center. "Back then, you'd spend hours just trying to repair the grille screen on a styled letter tractor, straightening wire where you could and welding in a patch in other areas. Today, you can buy replacement screen for any of the John Deere models. Plus, you can purchase new fenders, dash panels, and even hoods for a number of models."

Before you spend a lot of money on new parts, though, spend some time searching the salvage yards and used parts dealers for original parts that can be refurbished. The salvage yards are full of tractors that still have plenty of good parts on them, yet have damage that makes them cost-prohibitive to fully restore. The swap meets held in conjunction with a number of tractor shows are a good source of used parts, too. Not only will used parts make your tractor more original, but they also may save you some money. Just don't rely on used parts in critical areas where a failure could jeopardize safety or cost you more money later on.

While disassembling the tractor and cleaning parts, make a list of parts you'll need, so you'll have time to search parts sources and swap meets.

You'll be able to find everything from original parts that need work to reproduction parts and decals in the vendor section at many tractor shows.

Removing Broken or Damaged Bolts

There's a good chance that at some point during the restoration process, you're going to be faced with a bolt or fastener that simply can't be removed by ordinary means, such as a bolt that has broken off during removal or had the head stripped to the point you can't get a wrench on it.

One option is to drill a hole in the bolt, or what remains of it, and use an "easy-out" to back it out of the hole. Unfortunately, many restorers say easy-outs are more trouble than they are worth or cause more damage than if you just drilled the bolt out in the first place. "Easy-outs aren't," says one tractor restorer. "If you break one off, you have just created a bigger problem than you started with. The easy-out is made of some pretty hard steel, so once you've broken one off in place, you can no longer drill it out."

If part of the old bolt is still sticking above the surface, one alternative is to find a nut that is approximately the size of the broken bolt head. Place the nut over the broken bolt and then use an arc welder to fill the nut with weld, thereby fastening the nut to the bolt from the inside. In effect, you've created a new bolt head.

One of the last options, and the only one that is really effective on a bolt that can't be removed any other way, is to drill out the old bolt. Start by making a punch mark in the center of the bolt or bolt head. This will allow you to drill a starter hole through the center axis of the bolt. Begin drilling out the bolt using successively larger bits until the hole through the center is nearly as wide as the bolt. Be careful not to damage the threads in the parent material by using too large of a bit or drilling through a hole that is off center. In some respects, this process is like that used by a dentist doing a root canal. (Nice thought, huh?)

One restorer says he has had good luck using a set of left-handed drill bits to drill out a bolt, once a pilot hole has been established. Since the bit is turning in the direction you want the bolt to move, it will often come out as the heat increases and the center is hollowed out by the bit.

Above, both photos: **Sometimes, the only way to get a rusted bolt loose is to carefully apply heat and use a big wrench.**

Shaft Repair

One of the things you'll deal with most often in a tractor restoration is the replacement of seals, bushings, and bearings. Unfortunately, you may also run across the occasional shaft that has been damaged by a defective seal. This is generally evidenced by a groove in the shaft that is deep enough to feel when you run your fingernail across it. Replacing the seal at this point is not going to solve the problem. As long as the shaft is grooved, it's still going to leak.

The good news is you have a couple of options short of buying a new shaft. One is to sleeve the shaft with a sleeve, such as a Speedi-Sleeve, that fits over the original shaft to create a bridge over the groove. The Speedi-Sleeve is marketed by Chicago Rawhide, or CR for short, and is just one of several brands on the market.

To sleeve a shaft, you'll need to accurately measure the diameter of the shaft at a point where it hasn't been worn. Then, it's simply a matter of taking the measurement and the application information to your bearing or parts supplier. In most cases, the sleeve is thin enough that a different-sized bearing or seal is not required. The trick is just getting the sleeve installed on the shaft, using a special tool, since the fit is designed to be tight.

Another technique for repairing a groove or nick in a shaft is to fill it with a metal-type epoxy, such as Liquid Steel or J-B Weld. This is a viable option only if the shaft is not going to be subjected to fuel or lubricants that can act as a solvent over time. When using an epoxy, it's generally better to apply two or three thin coats and build it up, rather than one thick coat. After the compound has cured, the shaft can be sanded down until the repaired area is flush with the rest of the shaft. Just be sure to use fine-grain sandpaper or emery cloth to start, so you don't scratch the shaft. Finish it off by sanding the new surface with even finer-grit paper, in the neighborhood of 600-grit, for a smooth surface for the seal or bushing.

A final option, particularly if you're dealing with a large shaft, such as a rear axle or any shaft that receives heavy duty, is to take the piece to a machinist who can build up the area with a welder and then turn it down on a lathe.

There are a number of materials on the market, including brands like Liquid Steel and J-B Weld, that can be used to repair metal parts that aren't subject to a lot of pressure.

A number of parts, like the steering shaft, require hand sanding, due to their shape. Any part that can bind or stick should be refinished to like-new condition.

Chapter 5

Troubleshooting

In many respects, troubleshooting should be just a supplement to your tractor-buying procedure. In other words, there are several steps that could be considered troubleshooting that you should have already done in the process of evaluating the tractor when you bought it. However, for the sake of finding out how serious the problems are, let's take a look at a few more things.

Whether you are an experienced tractor mechanic or a novice working on your first tractor, your senses can tell you a lot about what is wrong—if you know what you are looking for. Your sense of smell, for example, can tip you off to a problem with the radiator, clutch, or engine. Your sense of hearing can tell a distinct difference between a tick, a knock, and a grinding noise. And there can be dramatic differences in the causes of such noises.

Start by looking on and under the tractor for traces of fluid. If what you find is a lighter hue, has the consistency of light maple syrup, and lacks the burned smell of combustion, it is probably from a leaking hydraulic fitting. A darker shade of brown that collects under the oil pan should be obvious. Engine oil is somewhat thicker and has the characteristic smell of having been in an engine; often it will leak out of the engine seals or leaky oil pan gaskets.

Whether you start the engine or just turn it over, you can check for spark at the spark plugs.

Evaluating a Tractor That Runs

Scott Carlson, service manager of John Deere Collectors Center, which restores all the tractors for Deere & Company, says that unless the engine is totally locked up, he likes to start up any tractor he receives and see how it runs in order to best figure out what it needs. If that means overhauling the carburetor or the magneto first, or installing a temporary gas tank, that's what he will do. At least then, he can listen to the engine, run it through the gears to see how the transmission and final drive sound, and evaluate the various systems.

Still, Chris Pratt, with *Yesterday's Tractors* on-line magazine, insists there are a few things you can check to avoid surprises later on. The following inspection points are just part of his troubleshooting and evaluation procedure.

1. Cooling System Inspection

Whether you're checking out a tractor before or after you've made a purchase, it's understandable that you want to fire it up without further delay. However, it will be to your advantage to check out a few things while the engine is still cold. This includes the cooling system, oil pan, and transmission.

First, carefully remove the radiator cap and check for coolant. There should at least be some type of liquid in the radiator. The best thing you can find is an antifreeze solution. The next best thing is clear water. If you find only clear water, though, you have to wonder if it was added since last winter or if it has been in there awhile.

If you find rusty water, you can expect to find pitting inside the engine and possibly a radiator that is leaking or about to leak. Rust in the cooling system is often an indicator that the coolant has become acidic, which means it could also start attacking metal components.

Last, but certainly not least, you'll want to make sure the coolant isn't oily. That can be an indication of seal failures; cracked parts, which are allowing oil and coolant to mix; or pitted parts, which can do the same thing.

If you find that the system is leaking, look for bad hose connections in the cooling circuit, especially where the hose ends meet the radiator or connect to the thermo-siphon components. Also check the radiator core for cracked tubing or leaky ends. If leaks in this area are excessive, the core should be replaced, since the internal integrity of the core itself is probably not worth salvaging.

From the radiator, you'll want to move back to the water pump, if the engine is so equipped. Look for a steady but slow dripping or for antifreeze streaks down the front of the engine housing. Most water pumps have a hole at the base that will leak antifreeze and coolant if the seal on the pump is in need of replacement.

2. Oil, Transmission Fluid, and Hydraulic Fluid Inspection

Next, you should take a look at the oil. But don't just pull the dipstick to see if the level is where it should be. Take a wrench and carefully loosen the drain plug to the point you could pull it out if you weren't holding it in place. Now, back off just enough pressure to drain out about a cup or less of oil and check it for water and antifreeze. If you get pure oil, you can rest a little more comfortably, knowing that water will settle to the bottom of a cold oil pan or gear case. A small amount of water can simply be condensation and may or may not be cause for concern. But if you find antifreeze, it should raise a red flag concerning the mechanical condition of the engine.

Repeat this process for the transmission and hydraulic reservoir.

3. Engine Starting Evaluation

Okay, now you're ready to start the engine. Does the engine start easily when cold? Knowing that a tractor starts after just a few revolutions easily eliminates many of your concerns in one check. You may not have a guarantee on the condition of each component, but you immediately know that the battery, compression, ignition wiring, distributor/magneto, fuel flow, and carburetor are in reasonable condition.

If the engine doesn't start easily, you may still be looking at a good tractor, but you know there is a little more work ahead.

Unfortunately, if you're looking at the tractor for the first time before buying it, and the unit has already been removed from the shed and warmed up prior to your arrival, you've lost out on a piece of information—namely the cold start.

4. Engine Running Evaluation

Does the engine run well when it is hot? Taking the time to see how the engine runs after it has been warmed up is particularly important if you are going to be using the tractor to pull a load. Plan to spend at least a half hour running the engine to check for problems that can cause it to run poorly after it warms up. In the meantime, you can look for leaks in both the engine and radiator.

After completing your inspection, shut the engine off and see how easily it starts when it is warm.

5. Exhaust Evaluation

Check for smoke from the exhaust. Blue smoke often indicates internal problems, such as problems with rings, pistons, or valve guides that are difficult and costly to repair. White or black smoke can frequently be corrected with carburetion or ignition changes.

6. Engine Noise Evaluation

Listen for noises from the engine. A ticking from the top of the engine may indicate the need for a simple valve adjustment, while a clunking or thumping sound deeper in the engine could indicate serious or expensive repairs. If possible, check if the sound becomes more pronounced under load. If a clunk becomes louder when the engine is put under a load, it may be an indication of problems with the crankshaft, bearings, or piston rods.

7. Oil Leakage Inspection

After the engine has been running for a while, shut it off and check the oil for foaming or the presence of water. Either condition is cause for concern. Check for oil seepage from the head and look for structural cracks in the block. These procedures may be difficult to perform if the engine is encrusted in dirt and grease, but they can be time well spent. Check over the cast and steel components and look for hairline cracks.

8. Electrical System Inspection

There should be some charge showing on the ammeter when the engine is running. You should also see some change in the charging level when the lights are turned on. This indicates that the regulator or resistor switch and cutout are operating properly.

9. Hydraulic System Inspection

If the tractor is so equipped, test the hydraulic system. Extend the rams on any hydraulic cylinder to test the range. If possible, use the system to lift a load, then let the load sit in the hold position for a few minutes to see if there is any leakdown. Chattering noises from the pump while lifting the load may indicate that the pump is getting an insufficient flow of hydraulic fluid. The pump will have experienced excessive wear if it has been run this way for long periods of time, and it may be on the verge of failure.

Simple Fixes

On occasion, a simple fix is all that is needed to correct what may seem to be a complex or an expensive problem. Always check the simple things first to avoid spending time and money on restoration steps that may not be necessary. Through his experience with tractor restoration, Chris Pratt has come up with the following list of common problems and simple fixes for them.

Runs Poorly When Warmed Up

Before replacing the carburetor, check the fuel line, sediment bowl, and tank outlet. With old tractors, these often become clogged with rust sediment and cause the engine to run as if the float and float valve are damaged. Quite often the tractor will run fine when started, but begin to starve out and miss after awhile.

Dies When Warmed Up

If the tractor warms up, then suddenly dies with no spark, and you find that the spark does not come back until the tractor cools down, the problem is most commonly a bad condenser. Since testing condensers seems to be a lost art, it is simplest to replace them.

Good Battery Won't Actuate Starter

This problem may be most common on tractors with a 6-volt system. Before replacing the starter, check for warmth at the connections of the battery cables. It may be that the cables are of too high of gauge (the wire is too small) or the connections may be less than perfect. As a rule, 6-volt systems draw more amperage than a 12-volt system, and the connections and wiring need to be near perfect for the starter to function as it was intended.

The Engine Is Getting Gas and Spark, but Won't Start

If the engine is getting a spark at the right time and gas is getting to the plugs, yet the tractor won't start, it is likely that your gas has gone bad—particularly if the tractor has been sitting for some time. The solution may be as simple as draining and replacing the gas.

Won't Start, Water in Distributor Cap

If you have trouble with your tractor during high-moisture times, such as during a thaw or in damp conditions, check under the distributor cap for moisture. In most cases, all you have to do is dry it out and hit the starter. Since it displaces moisture on electrical connections, WD-40 sprayed on the inside of the distributor cap can also do the trick.

Overheating or Not Charging

Before you replace your water pump, thermostat, and radiator cap, be sure the drive belt is the correct width and profile. Also ensure that it is tensioned properly. These factors can also cause the charging system to appear to be faulty.

Boiling Out Radiator Fluid

If the tractor is boiling out radiator fluid every time it warms up, the first thing you should do before replacing the thermostat is make sure the radiator cap is rated correctly for the system and that its spring and seal are still in good shape. A faulty cap may be letting off steam under what was supposed to be normal pressure.

Burning Oil

What if a compression test indicated good compression on all cylinders, showing that the valves, pistons, and rings are in good shape, but traces of oil smoke are coming out of the exhaust? This can be caused by the oil-bath air filter. Be sure that you are running the correct weight of oil. If the oil is too light, it will be drawn into the engine. Don't go overboard the other way, however; if the oil is too heavy, it won't clean the air. To learn more about oil-bath air filters, refer to chapter 13, "Fuel System."

Compression Testing

If you've had a chance to start or drive the tractor, you probably have an idea how well the engine runs. But for a real test, before you make the purchase or start tearing it down, you should run a compression test. This test measures the pressure built up in each cylinder and helps you assess the general cylinder and valve condition. It can also warn you of developing problems inside the engine.

Before you begin, you should start the engine and let it warm up to normal operating temperature. Then, shut off the engine and open the choke and throttle all the way to provide unrestricted air passage into the intake manifold. Remove all of the spark plugs and connect a compression gauge to the left-hand cylinder following the gauge manufacturer's instructions.

At this point, you need to either have someone else crank the engine or use a remote starter switch that has been connected to the starter relay. Always follow all manufacturers' safety instructions; make sure the transmission is in neutral and the wheels are blocked or locked prior to engaging the starter.

Crank the engine at least five compression strokes or until there is no further increase in compression shown on the gauge. Remove the tester and record the reading before moving on to the other cylinder. In addition to the psi rating, you should note whether the needle goes up all at once, in jerks, or a little at a time.

You'll need to check the service manual for your tractor for the recommended pressure. Engine compression specs can vary anywhere from 80 to 150 psi. Generally the pressure reading from one cylinder should be within 10 to 15 psi of the other. A greater difference indicates worn or broken rings, leaking or sticking valves, or a combination of problems.

If the initial compression test suggests a problem, you might want to confirm your suspicions with a "wet" compression test. This is done in the same way as your previous test, except that a small amount of heavyweight engine oil is poured into the cylinder through the spark plug hole before the test. Since this will help seal the rings from the top, it should help pinpoint the problem.

If there is little difference between the wet and dry tests, the trouble is probably due to leaking or sticking valves or a broken piston ring. However, if the wet compression reading is significantly greater than your first reading, you can assume the problem is worn or broken piston rings.

Your notes on how the needle moved up can tell you a lot, too. If the needle action came up only a small amount on the first stoke and a little more on succeeding strokes, ending up with a very low reading, burned, warped, or sticky valves are indicated. A low pressure build up on the first stroke, with a gradual build up on succeeding strokes, leading to a moderate reading can mean worn, stuck, or scored rings.

There can be good news, however. If the readings from both cylinders are within reasonably close proximity, you can assume that the upper end of the engine is in good condition and may not warrant an overhaul. A simple tune-up may suffice.

If you're checking the compression on a diesel engine, the process is basically the same, except you will need to remove the injectors and the seal washers. Plus, since the compression is higher on a diesel engine, you'll need to use either a different compression gauge or an adapter for diesel engines.

Squirting a little oil in the cylinder and performing a "wet" compression check will give you an idea whether a compression problem is caused by faulty valves or worn or broken piston rings.

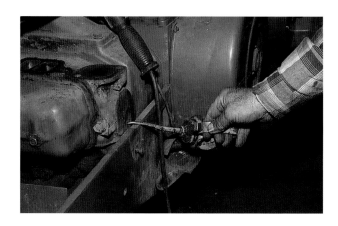

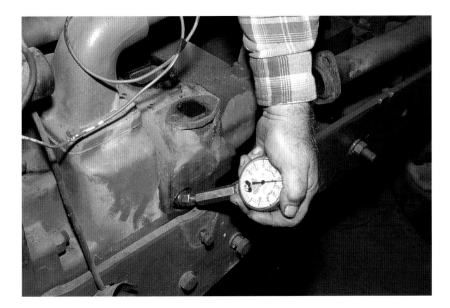

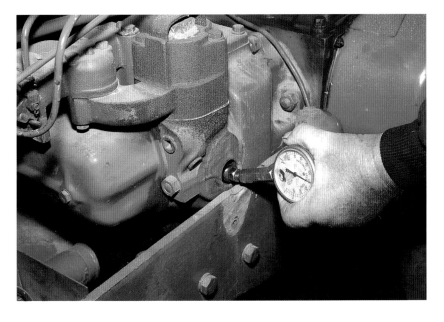

A compression check measures the pressure built up in each cylinder and helps assess the general cylinder and valve condition. It can also warn you of developing problems inside the engine. When performing a compression check, you'll want to record both the pressure reading and the rate at which the pressure increased. You should also compare the pressure readings between cylinders.

CHAPTER 6

Engine Repair and Rebuilding

The first real step in engine repair and restoration is to find out what you are dealing with and what repairs are necessary. Chapter 5, "Troubleshooting," should have given you some ideas about how much repair you might need to do. If you're lucky, you are restoring a tractor on which the engine is already in running condition. If that is the case, all that may be needed is a good tune-up.

On the other side of the coin is the engine that is completely frozen. Or the tractor may have been parked and left to rust after a piston rod broke or the engine block cracked. In either of those situations, you'd better figure on a complete engine rebuild.

The more common situation is the engine that will run or start, but performs very poorly. Perhaps it smokes or has a distinct knock. If you're like some restorers, you may choose to overhaul the engine as part of a restoration. And, if you're like others, you may be on a budget that demands just fixing what needs to be fixed. Either way, this chapter will guide you through some of the processes.

Above: This cut-away tractor, which was supposedly used by Deere & Company in sales demonstrations, reveals the simplicity of the John Deere horizontal two-cylinder engine. The unique tractor is currently owned by Lester, Kenny, and Harland Layher of Wood River, Nebraska.

Left: Some early kerosene-powered tractors utilized a water injection system that drew water from the engine water jacket in an effort to increase engine power. The other brass valve on the block is a compression release petcock, which partially releases compression on the cylinder when hand starting the engine.

Freeing a Stuck Engine

The first thing you have to understand about a stuck engine is that it took some time for the engine to set up and seize, and it will take some time to free it. Hence, freeing a stuck engine often takes more patience than anything.

Most often, engines get stuck because the pistons and cylinder walls or sleeves have rusted together. This can occur as a result of water entering the engine directly or by condensation—"block sweat"—inside the engine, which leads to flash rust. With a little time and work, flash rust can often be broken loose fairly easily. Pistons that are practically welded to the cylinder walls are a much bigger challenge.

Before you attempt to break anything loose, there are a few things to keep in mind. First, be very careful about towing the tractor in gear in an attempt to free the pistons. Even if you have soaked the pistons for some time, you run the risk of damaging the engine. Should the pistons, but not the valves, come loose, you risk bending the connecting rods or components in the valvetrain. You also need to be careful about putting too much pressure on any one piston—such as with a hydraulic press—if the pistons are still connected to the crankshaft. If the piston on which you're pushing comes loose, but the other one is still stuck, you risk damaging one or both of the pistons, as well as the connecting rod or the crankshaft.

One tractor restorer says he likes to soak the piston heads and cylinder walls while leaving the connecting rods attached and the tractor blocked up on one side and in gear. Then every few days, as he walks by the tractor, he gives the back wheel a push to see if anything has loosened up. Others have devised ways to put leverage on the flywheel or belt pulley. If you do attempt to tow the tractor at this point, make sure you're using a slow speed and you're pulling it over soft ground or gravel so the wheels will skid. Letting the wheels get a firm grip is a sure way to bend something.

Soaking the cylinders and using a press or hydraulic ram to force them loose is another option. One of the benefits of the horizontal two-cylinder engine is the fact that you can remove the head, disconnect the piston from the crankshaft, and take the entire engine block, with the pistons still in it, to a press where more pressure can be applied.

Rex Miller, a tractor restorer who lives near Avenue City, Missouri, has developed his own way of breaking the pistons loose on a John Deere two-cylinder tractor. He starts by pouring as much diesel fuel as possible into the cylinder through the spark plug hole. Then he installs a spark plug that has a hole bored through the center of it, into which he has soldered a grease zerk. Miller then pumps the remainder of the cylinder full of grease, adding it to the diesel fuel that already fills most of the cylinder. As he continues to pump grease into a full cylinder, the grease gun generates several hundred pounds of pressure against the cylinder head.

Another John Deere restorer says he built a special wrench that engages the flywheel. While one pin fits into a hole in the flywheel, another grips the outside flywheel's outside circumference. If he can't break the piston loose by putting pressure on the wrench, he positions a floor jack against the wrench handle and literally lifts the front of the tractor. In all but the worst cases, the weight of the tractor on the wrench eventually breaks the piston loose, and the front end settles to the floor.

When it comes to the type of lubricant or penetrating oil to pour into the cylinders, everyone seems to have their favorite recipe. Some simply use diesel fuel, while others prefer something as exotic as olive oil. Still others prefer a mixture of ingredients that may include brake fluid, penetrating oil, automatic transmission fluid, kerosene, Hoppe's gun solvent, oil of wintergreen, Marvel Mystery Oil, and Rislone. One restorer claims to have freed the pistons on fifteen different engines with a mixture of one-third automatic transmission fluid, one-third kerosene, and one-third Marvel Mystery Oil. Another prefers to use Sea Foam, a solvent available in most automotive stores and used for everything from cleaning carburetors to lowering the gelling temperature of diesel fuel.

In the worst case situation, you may actually have to destroy the piston to get it out. If the situation is that bad, though, there's a good chance you would have to replace the piston anyway. Occasionally, drilling several holes in the top of the piston will be enough to relieve pressure on the cylinder bore. If that doesn't do it, you may have to simply break the top out of the piston.

Engine Repair and Rebuilding / 63

Above: It took another tractor linked to the belt pulley to get this old D started for the first time.

One advantage of John Deere two-cylinder engines is that the block can be removed from the tractor and placed in a press when a piston is frozen in place.

One innovative method used to free a stuck piston is to fill the cylinder with diesel fuel, then pump grease and the accompanying pressure into the cylinder through a modified spark plug and grease zerk.

Ring Job or Complete Overhaul?

If you have seen traces of blue smoke coming from the exhaust and the engine has been using quite a bit of oil, there's a good chance that you are in the market for a set of piston rings. Before you can know for sure, you will need to check the piston-to-sleeve tolerances and surfaces, plus make sure the valve guides are not sloppy. The valve guides can exhibit the same symptoms as worn out pistons and sleeves.

In most situations, it is not advisable to replace only the rings in an engine, because by the time you do the teardown and measurement of the components, you'll find something else that justifies the need for a complete rebuild. If you are planning to replace only the rings, however, you first need to verify all of the following:

- Bore of cylinder is not scored
- Piston is not scored or cracked
- Rings are not stuck to the cylinder wall(s)
- Bore is within tolerance throughout piston travel (up and down and across right angles around the bore)
- Piston ring grooves are within tolerance and not damaged

Engine Block Preparation

Before you do anything else with the block, you'll need to check it for hairline cracks. Depending upon the size of any cracks you find, you have a couple choices. One is to find a new block at the salvage yard. The other is to have the block welded by a competent welder.

One of the most common problems is the lack of flatness of the mating surface between the head and block. To check both surfaces, you'll need a straight edge, feeler gauges, and your tractor shop manual, which will provide the allowable tolerance. If the surfaces aren't flat, you'll need to have the errant surface milled by a competent machinist.

Cylinder Head and Piston Removal

If you have a service manual for your tractor, it is best to follow the disassembly process outlined for your particular model. In general, the process begins with draining all the fluids from the engine, including water, antifreeze, fuel, and oil. You'll also need to remove the engine hood, fuel tank, exhaust pipe, and lower water pipe, as well as the toolbox and generator on models so equipped. Finally, disconnect the fuel line and controls and remove the carburetor.

Next, remove the air inlet elbow, if applicable, along with any lines connected to the tappet lever cover. Then remove the tappet lever cover itself. With the cover off, turn the flywheel until you can see that all valves are closed and remove the tappet levers assembly and push rods. On early model A Series tractors, it's important to also remove the left tappet lever bracket retaining stud to prevent damage to the air cleaner.

Now, you can remove the nuts from the cylinder head retaining studs and slide the cylinder head forward and off the studs.

Once you have the head off the block, you can remove the ridge at the top of the cylinder, using a special tool called a ridge reamer. This ridge is formed naturally as the piston and ring travel up and down thousands of times in the cylinder. In essence, the area of piston travel wears, while the portion above it does not. While it's not absolutely necessary to remove the ridge on some engine overhauls, it's more or less imperative on a horizontal two-cylinder engine, since there's only one way to remove the pistons. Since the ridge reamer is one of those tools that is only used on occasion, perhaps you can borrow or rent one. Some restorers simply use an air-powered grinder or Dremel to remove the ridge.

Now, you can remove the rod-bearing caps and pull the pistons out the front of the block. Examine both the pistons and the cylinder walls for scoring that would suggest the need for more than just ring replacement. In the process, try to keep carbon from getting into the cylinders, particularly the water and oil passages.

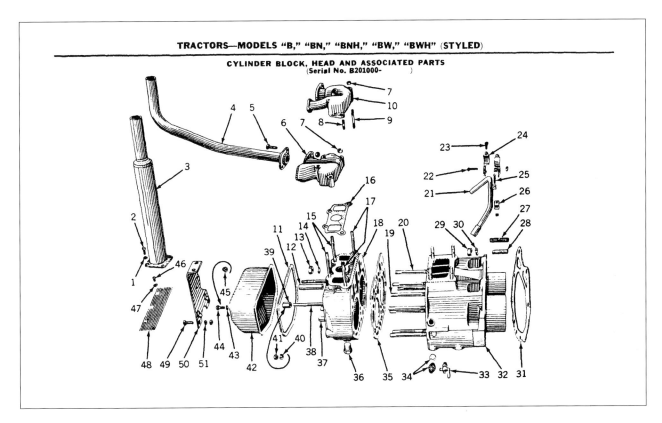

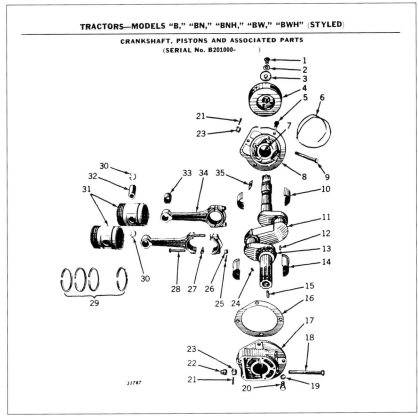

Above: This parts book drawing illustrates the cylinder block, head, and related parts.

Left: A second drawing is used to show the orientation of pistons, crankshaft, bearings, housing, and the clutch drive disk.

Above: For a complete overhaul, the engine may need to be removed from the frame. Or the tractor will need to be split if the engine block serves as a load-bearing member.

Above, both photos: Since the block rests within the frame, it's fairly easy to work on most Waterloo-built John Deere two-cylinder engines. For a simple ring job, all you have to do is remove the head.

The vertical cylinder engines used in Dubuque-built two-cylinder tractors call for an entirely different overhaul procedure.

Before reinstalling the head, it's a good idea to hit the mating surfaces with a sheet of fine sandpaper or emery cloth.

Savannah, Missouri, restorer Roy Ritter decided to pull the block on one two-cylinder so he could better remove rust from the water passages.

The fact that this two-cylinder John Deere block had rusted through where it contacted the frame wasn't noticeable until the block was pulled.

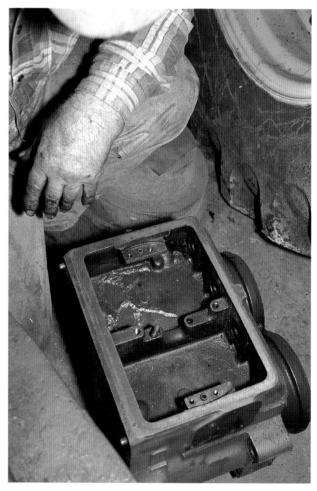

At some point, a previous owner had decided to have this two-cylinder block welded. Rather than reuse it, Estel Theis chose to replace it with a used block.

Piston Ring Replacement

Once you have removed the pistons and checked them for scoring, you'll need to remove the rings by carefully spreading them from the break or ring gap. This is most easily done with a special tool called a ring spreader. An inexpensive ring spreader looks like a pair of pliers that open when squeezed. More-expensive ring spreaders have the same design, but they also have a band to wrap completely around the circumference of the ring to ensure that you don't elongate or spread the ring gap too far. This is not that important on your old rings since you are throwing them away, but on the new ones, it is critical.

With the rings removed from the pistons, examine the grooves that the rings fit into and make sure they are not damaged. Carefully clean the grooves to remove carbon and dirt that would hamper the correct seating of the new rings.

Next, you'll need to determine the size of the new rings by measuring the bore and determining what oversize will completely fill the gap when the piston is at the top of its stroke. The manufacturer's required ring gap should also be taken into account. The ring gap is the clearance left at the split in the ring when the ring is as compressed as it will be in the cylinder. This usually occurs at the top of the piston's stroke.

In most cases, going one size up from the existing rings is sufficient, since ring replacement is done only when there is little wear on the piston and cylinder. If it takes more than one size increase, you might want to consider a more thorough overhaul.

Once you've determined the ring size, you'll need to hone the cylinder to remove the smoothness—generally referred to as a *glaze*—from the cylinder bore. Otherwise, the new rings will not seat properly. A cylinder hone that fits into a ¼-inch drill can be found at most any auto parts store.

The goal of cylinder honing is to get a nice crosshatch surface on the cylinder. This requires moving the hone up and down as the drill operates. Never allow the drill to run in one spot, and keep the hone lubricated and cooled with a fifty-fifty mixture of diesel fuel and another lubricant, such as kerosene or penetrating oil. Be sure there are no large particles on the bore or hone surfaces that will cause scoring. If you're working on a vertical cylinder model, cover the crankshaft rod journals while honing to keep them protected from falling debris.

Now, it is time to check the ring end gap. As stated earlier, the ring has to have the minimum specified compressed gap when it is in the cylinder bore to allow for expansion that occurs when the engine reaches operating temperature. Otherwise, the ring ends might butt together and cause scoring and ring breakage. On Deere's two-cylinder engines, it's generally in the range of 0.004 to 0.005 per inch of cylinder bore diameter, but check your repair manual for the exact specification. Note, too, that the ring end gap often differs between the compression rings and the oil rings.

To measure the gap, you'll need to compress the first ring and place it inside the bore. Don't put it on the piston. Now, push the ring into the cylinder using an inverted piston. This not only makes it easier to push the ring into the cylinder, but it also ensures that the ring is square with the cylinder wall. Take your feeler gauge and measure clearance between the ends of the ring. Compare this measurement with the specifications in your manual and determine what changes, if any, are necessary.

Insufficient clearance will require that the ring gap ends be filed down to tolerances. One of the best ways to do this is to take a file and mount it vertically in a vise. Take the ring and, holding it firmly in both hands, draw it downward over the stationary file. After removing a small amount of metal, check the ring in the bore again, repeating this process as often as needed. Follow this procedure for each ring, making sure you note the gap specification for each particular ring.

After the gap for each ring has been established, take each ring, insert the edge of it into the corresponding ring groove in the piston, and measure the side clearance to determine if the ring grooves are worn. There should be ample room for the proper feeler gauge between the ring and piston lands. This is another case where you need to check the manual for the specifications. Too tight a fit will keep the rings from rotating and moving properly as the piston moves up and down. Excessive side clearance, on the other hand, will allow the rings to flutter in the piston grooves when the engine is running. This can result in poor sealing of the combustion chamber or, worse, eventual ring breakage. Make sure you place each ring in corresponding order with each piston groove and check the ring for a mark indicating the correct side up. If the ring side clearance exceeds specifications, you'll need to replace the piston.

Once everything has checked out and the ring gap has been established, it's time to install the new rings on the piston using the ring spreader to prevent overexpansion or distortion. To be honest, you may not have a ring spreader large enough for the job if you're restoring something like a D, G, R, or their successors, all of which boast pistons larger than 6 inches in diameter. The pistons on the D are, in fact, a whopping 6.75 inches in diameter. If you're not using a ring spreader, very carefully spread the rings by hand and slip them into the ring grooves starting with the lowest ring (the oil ring) and ending with the top ring (the compression ring). Be sure to stagger the piston ring end gaps around the piston for maximum sealing.

Now, it's time to put the pistons back into the cylinder. Remember, all components—especially the pistons—should be reinstalled in their original positions if they are being reused.

First, lubricate the pistons and cylinder walls with clean engine oil. Then, using a suitable ring compressor to compress the piston rings, place the piston into the bore. Basically, a ring compressor is a sleeve that fits around the piston to compress the rings enough to allow the entire piston to be slipped into the bore. Unfortunately, you may again have problems, due to the monstrous size of some two-cylinder pistons. The stroke on the R and its successors, for example, is 8 inches. Few ring compressors will cover that length. One trick Kansas restorer Dennis Funk uses is to cover the top rings with the ring compressor, while compressing the bottom oil ring with two large screw-type hose clamps linked together. Be sure the ring compressor is perfectly clean on the inside. Gently tap the piston into the cylinder using a soft block of wood or the handle from your hammer. Be careful to ensure that the connecting rod studs don't scratch the cylinder walls or the crankshaft journal as you're installing the pistons. As extra insurance, you can always place pieces of plastic or rubber tubing over the connecting rod studs during piston installation.

At this point, all that is really left is to apply a light coat of engine oil to the connecting rod bearing inserts and install the bearing caps on the connecting rods.

Before you reassemble the engine, you should check for excessive end play between the camshaft and its bushings to see if the bearings have exceeded their useful life. Since the connecting rod bearing caps are already off, it's a good idea to also perform measurements to see if the rod bearings or inserts need adjustment or replacement. On many machines, adjustment will involve simply removing one or more of the shims, interchanging shims, or, in the case of later models, installing new bearing shells. If you don't do it while you're replacing the rings, you may have to go through the whole process again in the near future. For more specifics on determining serviceability of these components, refer to the appropriate sections, which follow.

Carefully remove the old piston rings without damaging the piston or the grooves. A ring spreader makes the job a little easier.

To determine how much, if any, the ring grooves have worn, insert the new piston ring in the appropriate groove and measure the side clearance. Your repair manual should list the allowable tolerance.

Before installing new piston rings, carefully clean the ring grooves to remove carbon build up or residue.

Right: As part of the ring replacement process, you'll need to determine the size of the new rings. This is done by measuring the bore and determining what oversize will completely fill the gap when the piston is at the top of its stroke.

Above: While honing the cylinder, keep the drill moving to create a nice cross-hatch surface, and keep the hone well lubricated.

Right: Before installing new rings on a piston, squarely position each ring in its respective cylinder, measure the end gap as shown, and compare it to the engine specs.

Left: Insert the base of the piston into the bore and compress the rings with an appropriately sized ring compressor.

It takes a large ring compressor to go around the piston on a Model D or Model R. Both have a bore that exceeds six inches.

With the rings compressed with a piston ring compressor, and the piston lined up with the bore, carefully drive the piston into the bore using a soft tool or the end of a wooden hammer handle.

Main and Connecting Rod Bearings

The main and connecting rod bearings can generally be lumped together, since the methods used to measure them are identical. The measurement you are looking for here is the existing size of the crankshaft journal. Using this figure, you can follow the appropriate procedure for bringing the crank or rod journal tolerance back to factory specification.

The first step in measuring the crankshaft and bearings is to check the surfaces for scoring. If scoring is minimal, it can generally be remedied by having a machine shop turn the crank on a lathe; however, this will cause the journal to be undersized to its original specification. If the scoring or damage is too bad, you'll simply have to look for a replacement crankshaft.

Assuming the crankshaft passes the initial inspection, use a caliper to determine if it has equal wear across the surface. If the journal has more wear on one side than the other, you will again have to look at having it turned or replaced. Finally, check to see if it has worn unevenly around the diameter of the journal, creating an oblong cross section. This can be done by taking a measurement, rotating the caliper around a quarter turn, and taking another measurement. Repeat this process a few times and you will quickly get an idea whether it is serviceable. The better manufacturers' manuals will even explain what are acceptable tolerances in this respect.

If the crankshaft does exceed wear tolerances, you have two options. The first is to purchase a reground crankshaft. While this is expensive, it has the benefit of simplifying your measurements, since new bearings that have been pre-measured for the shaft are generally provided with the replacement. In this case, you are done with this job.

The less-expensive alternative is to have the crankshaft ground by a local machine shop. While the price is generally reasonable, there is often a wait, especially during certain times of the year, since many of these shops also work on specialty engines, such as those used for racing or tractor pulling.

Before you take a crankshaft in to be ground, be sure you can get the right-sized bearings to cover the amount of material the machinist will remove before you have the shaft ground. Otherwise, you may have wasted your money on the machining.

The other journal- and bearing-measurement process is accomplished by putting the whole assembly back together with Plastigage inserted between the bearing shell and the crank journal.

You should be able to find Plastigage at any good auto parts store. You'll find that it comes in different colors, like red, green, and blue. The color is a universal code for the range of clearance each particular plastic thread is capable of measuring. For example, red Plastigage is designed to measure a bearing clearance of 0.001 to 0.004 inches. If you take along your service manual, or tell the parts person how you want to use it and how much clearance you need, you shouldn't have to worry about colors. Any knowledgeable parts salesperson should be able to help you find the right size.

Basically, Plastigage is an impregnated string that squishes flat as it is squeezed between the journal and bearing shell. To use it correctly, cut a strip of Plastigage wide enough to go across the journal; then reassemble the shell and cap with the strip placed between the journal and shell; and torque the bolts to the proper specifications. Never turn the crank during this process.

Next, remove the bearing shell and compare the width of the Plastigage against a scale on the package to determine your exact clearance. Once you find this value, you can determine how much oversize will be required to bring the clearance back to that required by your manual.

As you've probably noticed, you have now measured the journals twice, once with a caliper and once with Plastigage. But with many old tractors, this is important. The caliper finds the irregularities and gross undersizes that necessitate crank welding, grinding, or total replacement. The Plastigage process is needed to determine whether shimming is required during reassembly. If Plastigage is the only measurement you use, it may be hard to spot irregularities like conical or oblong journals. On the other hand, you virtually can't measure for the proper clearance without Plastigage.

On a number of Deere models, including the A, B, G, and H, the main bearing bushings are not available as individual parts, but only as a unit with the main bearing housing, on a factory-exchange basis. On early production A tractors, field installation of the sleeve-type bearing and housing units can be done, provided the crankshaft is not excessively worn. On Model D tractors, the main bearing clearance can be adjusted to a tolerance of 0.002 to 0.006 by varying the number of shims. Finally, on Waterloo-built 20 and 30 Series tractors, the sleeve-type main bearings are available as individual repair parts, which must be sized after installation in the main bearing housings. They're also available on a factory-exchange basis as part of the main bearing housing.

The connecting rod bearings vary between models and years. Early production A, B, and G models use rod bearings that are of the shimmed type, which means they can be adjusted by varying the number of shims between the rod and the cap. Later models built in both Waterloo and Dubuque were equipped with slip-in, precision-type rod bearing inserts, which continued to be used through the 30 Series. Connecting rod bearing inserts are available in standard as well as various undersizes, depending upon the model. In almost all cases, the undersize choices include 0.002 and 0.004 on Waterloo-built models and 0.003 and 0.005 on Dubuque-built models. In some cases, older-type bearings can be replaced with precision-type rod bearing inserts, provided the new-type connecting rods are also installed.

Early D models, on the other hand, used bronze-backed, babbitt-lined bearings that can be shim adjusted; later models used spun, babbitt-type shimmed bearings.

In all cases, the actual adjustment is best confirmed by checking the clearance with Plastigage thread once the bearings have been shimmed or replaced.

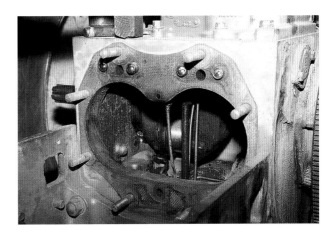

Part of engine overhaul includes checking the condition of the crankshaft.

The stripped flywheel splines on this two-cylinder crankshaft have essentially rendered it useless.

Connecting rod and main bearing clearance can be easily checked for tolerance using the appropriate size of Plastigage.

Most connecting rod bearings are the slip-in, precision type and are available in both standard and various undersized sets.

When tightening the connecting rod cap bolts, refer to your service manual for the correct torque rating.

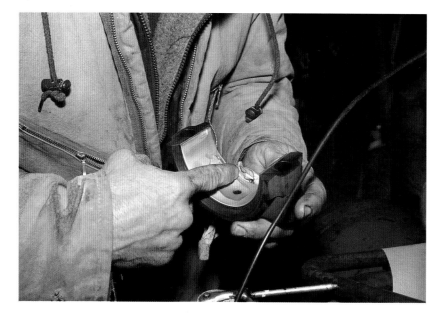

When installing new bearing shells, make sure the shell projections engage the milled slot in the connecting rod and bearing cap and the shell has been well lubricated.

Camshaft Overhaul

Although camshafts can bend, it's not likely, since the push rods tend to sacrifice themselves much sooner. The more common problem with the camshaft is worn or scored bearings, which calls for bearing replacement. This can result when the surface went too long without oil or if a foreign object lodged in between the bushings and camshaft.

While it is possible to get the shaft turned, you'll need to first check to see if oversize bearings are available for your model. Considering the availability of John Deere parts, the simplest solution may be to purchase a used or reconditioned camshaft and stick to standard-sized bearings.

Another likely problem you'll find with the camshaft is excessive end play. It's important to note that there is a substantial difference between the amount of camshaft end play allowed on older two-cylinder tractors and newer ones. On A and G Series tractors, for example, the camshaft end play is controlled by a thrust spring, which is located behind the camshaft right bearing cup. On B and D Series tractors, the recommended camshaft end play of 1/64 to 1/32 inch is achieved by varying the number of shim gaskets under the left bearing. The 20 and 30 Series Waterloo tractors, meanwhile, call for a desired end play of 0.005 to 0.012 inch, which must be measured with a dial indicator. As with the B and D Series, the end play is controlled by shims located under the bearing cap. As always, refer to your service manual for allowable tolerance and method of adjustment.

When reinstalling the camshaft, be sure to mesh the camshaft gear with the crankshaft gear as instructed in your service manual to ensure that the timing marks are in register. On a number of models, there's also a mark on the camshaft to ensure that the camshaft gear is reinstalled correctly and in register.

It doesn't take a measurement to determine that these camshafts are unusable. Missouri restorer Roy Ritter says that much of the damage was caused by mouse urine, which tends to be highly corrosive.

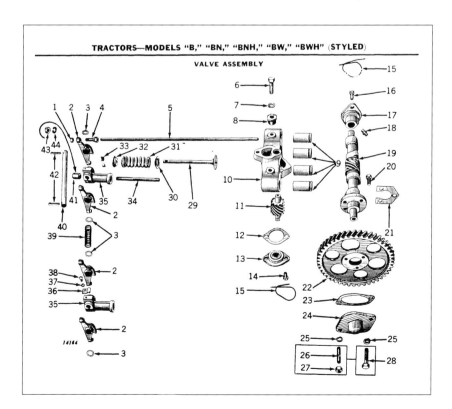

Although it applies specifically to the Model B, this illustration shows the relationship of the valve assembly components found on the majority of Waterloo-built models.

Push Rods

Although it's not likely, push rods can bend if the valve timing was off or even if a valve was adjusted to be open all the time. Visibly bent push rods should be replaced and all others should be checked. Checking them can be done with a perfectly flat surface and feeler gauges. If they are not absolutely straight, it's usually best to replace them, even if it means going to a salvage yard for better ones.

Rocker Arms

Like the push rods, you'll need to check the rocker arm, or valve lever shaft, for straightness and a smooth profile. Rocker arms can be distorted, which can not only make adjustment difficult, but also cause the push rod to slip when combined with a slightly bent push rod. If the shaft is bent, you'll need to find a replacement. However, if you find that only the lobes are pitted, you can have the shaft reground.

Cam Followers

Waterloo-built two-cylinder engines use mushroom-type or barrel-type cam followers, depending upon the tractor model. These cam followers operate directly in the unbushed bores of the follower guide and do just as their name implies: They follow the cams on the camshaft and transfer the motion to the push rods. If the clearance between the followers and the case bores is excessive, the best option is to replace the defective parts.

Be sure to inspect the cam followers and push rods.

Valves and Valve Seats

Faulty valve action is one of the main reasons for loss of power in an engine. Although carbon, corrosion, wear, and misalignment are inevitable products of normal engine operation, the problems can be minimized with high-quality fuel and valve tune-ups.

Carbon is a byproduct of combustion, so it's always going to cause some problems, like fouling spark plugs, which make the engine miss and waste fuel. But valve seats can also become pitted or be held open by carbon particles. Carbon deposits can also insulate parts and cause them to retain heat, compared to clean metal, which tends to dissipate engine heat. Heat retention increases combustion-chamber temperature and causes warping and burning. Finally, unburned carbon residue can gum valve stems and cause them to stick in the guides.

Consequently, valves need to be carefully checked for misalignment, distorting, burning, pitting, warping, and out-of-round wear, especially on the exhaust valve, since it is exposed to the high temperatures of exhaust gases.

Burning and pitting are often caused by the valve failing to seat tightly due to carbon deposits on the valve seat. Poor valve seating may also be due to weak valve springs, insufficient tappet clearance, warpage, and misalignment. This, in turn, permits exhaust blowby, which can lead to warpage in the upper valve stem due to the exposure to intense heat.

Out-of-round wear follows when the seat is pounded by a valve whose head is not in line with the stem and guide. Oil and air are sucked past worn intake valve stems and guides into the combustion chamber, causing excessive oil consumption and carbon build-up.

While distortion is generally caused by unequal tightening of cylinder head bolts, misalignment is a product of wear or warpage, often hastened by poor lubrication. Both problems contribute to sloppy clearances and poor valve sealing.

Valve Guides

Valve guides tend to warp because of the variation in temperatures over their length. Consider, for example, that the lower part of the guide is near combustion heat, while the upper is cooled by water jackets.

Engine Repair and Rebuilding / 79

Although this two-cylinder head is salvageable, water damage has already ruined at least one valve.

Valves need to be carefully checked for warping, burning, pitting and out-of-round wear, even if they don't have rust damage like some of these.

Both photos: **A valve spring compressor is used to compress the valve springs for access to the keeper—a small pin, clip, or wedge that needs to be removed to take off the spring.**

Any wear, warpage, or distortion affecting the valve guides destroys their function of keeping the valve head concentric with its seat, which obviously prevents sealing. If the problem can't be remedied with valve grinding, it will be necessary to replace the valve.

Valve Springs

Valve springs must be of a uniform length to be serviceable. To check the springs, place them on a flat, level surface and use a straight-edge ruler to determine whether there is any irregularity in height. Unequal or cocked valve springs should be replaced.

Spring tension that is too weak will also allow the valves to flutter. This aggravates wear on the valve and seat, and can result in valve breakage. If the springs are less than 1/16 inch shorter when compared with a new one, they should be replaced.

Valvetrain Overhaul

Depending upon the condition of the engine when you started, you may or may not have to do major work on the valvetrain. If the engine was running when you acquired the tractor, a good visual inspection and adjustment of the tappets may be sufficient.

However, when water gets into an old engine, it seems that the valves are among the first components to suffer. In most cases, restoration can be accomplished by using a valve seat tool to regrind the seats and grinding or lapping the valves. However, if damage to one or more of the valve seats is severe, it may be necessary to have a machinist cut recesses into the block or head that will accept special hardened valve seats. These, in turn, can be ground to the correct seating angle for your engine. Some people like to install hardened seats as a routine, so the tractor can handle unleaded gas without fear of later damage.

To inspect, grind, or lap the valves and seats, you'll first need to remove them from the head or block, depending on whether it is an overhead-valve engine or a valve-in-block engine. (Only the earliest John Deere tractors, including the GP, have valve-in-block engines.) To do this, you'll need a valve-spring compressor so you can compress each of the valve springs. Once the spring is compressed, you can remove the split-cone keeper from the base of the valve spring. At this point, the valve can be lifted out. Be sure to keep the valves separate so each one can be replaced in the original seat. Keep in mind, too, that intake and exhaust valves are not interchangeable.

Depending upon the condition of the valves, you'll have one or more options. If the valves have notches or burned sections, you'll need to replace them. However, if the valve faces and seats are simply a little rough and dirty, they can be cleaned and renewed with valve lapping. To do this, make sure the valves are returned to their original position after inspection, so you're lapping each valve into its original seat.

Before you install the valve, though, place a small bead of medium-grit valve-lapping compound around the face of the valve where it mates to the valve seat. Now, using a valve-lapping tool, rotate the valve back and forth in the seat, alternating between clockwise and counterclockwise. In the simplest form, a valve-lapping tool is little more than a suction cup on a stick that you spin in your hands. Other, more sophisticated versions have a suction cup on a shaft driven by a device that looks a little like an egg beater.

At any rate, you can move on when you have a smooth, clean surface all the way around the valve face with a matching surface on the valve seat. If necessary, add more compound during the process.

Valve lapping should not be a substitute for having the valves professionally reground. If they're in such rough condition that a few minutes of lapping doesn't polish up the mating surfaces, you're probably better off having a machinist grind the valves and seats. This basically consists of clamping the valve stem into what looks like a large drill chuck and turning it against a grinding wheel that renews the face to the correct angle.

Unlike later Waterloo-built tractors, the Model GP has the valves in the block, rather than in the head.

All that is left now is to thoroughly clean all the parts, including the keepers. Then check the valve springs to make sure they meet service manual specifications and reassemble the valvetrain, coating all parts with clean engine oil as you go.

Once the valves and springs have been assembled back into the head or block, you'll need to finish up by adjusting tappet clearance according to specifications. Correct clearance contributes to quiet engine operation and long valve-seat life. Insufficient clearance causes the valve to ride open, resulting in lost compression and burning. Too much clearance retards timing and shortens valve life.

Above: To ensure the valve seats are evenly resurfaced, a guide is first installed in the seat.

Right: Resurfacing the valve seats with a valve seat grinder can improve valve seating and combustion to like-new condition.

Above: Valves that are slightly pitted or worn may need to have the face ground on a valve grinding machine.

Left: To lap the valves, coat the valve face with lapping compound, which contains fine abrasives, and place the valves in the appropriate valve guides (without any of the springs installed).

Right: Using the valve-lapping tool, which usually has a suction cup on one end, rotate the valve until the face and seat have matching polished areas that are smooth and clean.

Below: When properly finished, both the valve and the valve seat will have the type of smooth surface that ensures a tight seal.

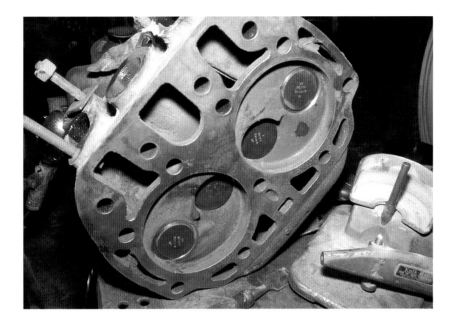

Having been cleaned and lapped, the valves are reinstalled in the head or block, depending upon the engine configuration.

The last step in valvetrain overhaul is reinstalling the valve springs and adjusting the tappet clearance to specifications.

Having been lapped and reinstalled, the valves in this head promise compression ratings rivaling those of a new engine.

Pistons and Bores

Just as you did with the crankshaft, you'll need to examine the pistons and cylinder bore for any visible damage, such as scoring, out-of-roundness, and greater width on the bore at the highest point of piston travel (not including the ridge that forms above the high point of travel). Scoring will indicate the need for replacement regardless of the measurements. The other conditions can be determined through precise measurement.

To check for out-of-roundness, use an inside micrometer to take a measurement at the top and bottom of the cylinder. Then repeat the process at right angles to the original measurements. With these numbers, you can see how conical the cylinder is and how oblong it has become. Repeat the process using your caliper on the piston. There will usually be a factory specification for what is acceptable.

The amount of deviation from the factory specifications on both the piston and the bore will determine whether or not the parts need to be replaced. If the cylinder bore is seriously scored or out of round, the only solution is to have the block bored out. That leaves you with one or two options, depending on price and severity. If the damage isn't severe, you may be able to purchase oversize pistons and install those in the bored cylinders. For most Deere models, cast-iron pistons are available only in standard and 0.045 oversize. For Model H tractors, the options are limited to standard and 0.030 oversize, while D owners can go up to 0.090 on oversize. Many machine shops, however, can fit the bores with aluminum or alloy pistons in larger sizes.

If the damage is too extensive, you'll need to install a set of sleeves to fit the old pistons. Depending on the price you have to pay for oversize pistons and rings, you may decide that you're money ahead to install sleeves anyway. If the pistons are in good shape, you may find that it's less expensive to install a set of sleeves that fit the new bore and your old pistons.

According to Scott Carlson, service manager at John Deere Collectors Center, there are some restorers who take other things, like sound and balance, into consideration. They claim that if anything other than a cast-iron piston is used, the resonance just isn't the same. Since the crankshaft and flywheel were designed for cast-iron pistons, the tractor isn't going to run quite the same with lighter-weight aluminum pistons. Consequently, many restorers would rather sleeve the engine and stick with standard cast-iron pistons than compromise with an alloy.

"We're seeing guys do that more and more all the time, which is making it a little hard to find original cast-iron pistons for some models," Carlson says.

If the differences at the top and bottom are acceptable, you may still need to determine if there is too much distance between the piston and the bore. This will be most noticeable at the top of the stroke. Most shop manuals suggest using a long feeler gauge of a certain size placed between the piston and bore or sleeve with the piston inserted in the cylinder. A scale is then used to pull the feeler gauge out of the bore. The amount of drag required to pull it free, as measured on the spring scale as pounds of pull, determines its serviceability. This process needs to be done without the rings in place, and the measurement is usually taken 90 degrees from the piston pin hole.

Piston Pins

The piston pin is a full-floating-type pin retained in the piston pin boss by snap rings. For most tractors, pins are available in standard, 0.003, and 0.005 oversize. In most cases, the pin should have a thumb-push fit in the unbushed piston and have a clearance of 0.0002 to 0.0024 in the connecting rod bushing, depending upon the model. As for diameter, it will be necessary to use a caliper and compare the reading with the diameter specifications listed in your repair manual.

It's important to check the fit of a piston pin as part of the engine overhaul process. Your manual should tell you how tight the pin should fit.

Rebuilding the Engine

If you're planning to completely overhaul the engine, you may want to consider purchasing an engine rebuild kit—assuming there is one available for your tractor. Most have all the replacement parts you'll need for overhauling an engine without going to the trouble to separately locate gaskets, special measuring tools, or miscellaneous parts. But don't let the matching pistons, rings, and bearings lead you into a false sense of security. They, too, need to be measured and installed in a workshop environment.

If you are going for a total rebuild, your best bet is to have the engine block, head, and any other major grease-carrying components professionally cleaned in a hot chemical bath. This process will not only strip those components of grease and paint, but it will also remove carbon, varnish, and oil sludge from both the inside and outside of each part.

While the engine components are out for cleaning, it's also a good idea to have a professional magnaflux the engine block and head for cracks that are not visible to the naked eye and, as a result, would have been overlooked in your original inspection. This investment alone could save you a lot of time and money.

Meanwhile, smaller parts can be cleaned in the shop using a solvent tank, then dried, inspected for integrity, and placed on a clean rag.

When the engine comes back from the machine shop, use an air hose from your air compressor to blow out the engine block. Be sure to hit all the openings and bolt holes to remove any residue that may be left. A clean solvent rag can be used to wipe off the internal surfaces to remove any film left over from the tanking process. Hopefully, the machine shop has reinstalled the casting plugs (sometimes called freeze plugs) and the oil galley plugs.

It is generally recommended that the head be reworked in its entirety by the machine shop, as well. Pressing in valve guides, getting the correct angles on the valve seats, setting the proper valve-to-head recession setting, and measuring and milling warped surfaces, among other things, are really beyond the scope of the average home shop level.

Crankshaft and camshaft measuring, grinding, and polishing are also out of the realm of the home shop. It is wise, though, to always double-check the machine shop's work with a feeler gauge, dial indicator, and Plastigage. It is important to check the crank end play with a feeler gauge and also to check each bearing journal with Plastigage before starting the reassembly procedure.

If your rebuild requires the use of cylinder sleeves, you may want to have the machine shop install these, as well, since it will be necessary to cut a recess into the block. If you're simply honing the cylinder walls, you can refer back to the piston ring replacement section for this process.

Once the cylinders have been honed or sleeves installed, the next step is to match the pistons to each cylinder. Refer to your shop manual and locate the piston-to-cylinder wall measurement. To obtain this measurement, you'll need to locate a long feeler gauge, commonly called a ribbon gauge. Install the piston in the bore and see if the specified feeler gauge will slip in next to the piston. The proper fit is when the ribbon gauge can be pulled from between the piston and sleeve with a specified inch-pound pull as verified by a reliable scale. If you don't have the recommended scales, you should at least make sure the feeler can be pulled out freely with moderate pressure and without binding.

Each piston will come with a certain number of piston rings. Unwrap the rings and lay them next to each piston in the order of installation. Basically, the instructions for measuring and installing the piston rings are the same as for piston ring replacement, so refer to the previous section in this chapter about that process.

Once the piston rings have been installed, the next step in assembly is to check the piston pin-to-bushing tolerances. Once again, this is a very close tolerance and is best done by a competent machine shop. An insertion pressure of pin to bushing will be slightly looser than that of the pin to piston. When dealing with clearances of 0.0002 inches or less, it is important to have the proper measuring tools. A pin too tight or too loose will cause a piston and rod to break apart under the stresses of engine operation.

Be sure to lubricate all parts during reassembly to prevent scoring or damage.

It's also a good idea to make sure the threads on the head bolts are clean and haven't been damaged during the overhaul.

Engine Repair and Rebuilding / 89

Before reinstalling the cylinder head, slip a new head gasket over the mounting bolts, making sure it is oriented correctly.

To further ensure against air and oil leaks on this old GP valve-in-block engine, Estel Theis makes washers out of strips of solder. When compressed by the head bolt nuts, the material will completely fill the cavities around and beneath the nuts.

When tightening the head bolts, be sure to alternate the tightening pattern and torque the bolts to spec.

Adding the valve cover should finish up this John Deere two-cylinder overhaul.

Don't forget to reinstall the fan drive shaft before installing the manifold(s). You may have to retrace a few steps.

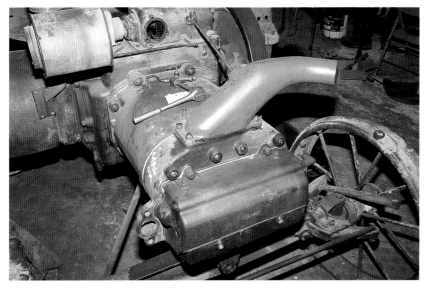

There's still a lot of work left on this old Model D, but at least the engine is finished.

Having overhauled the engine, Ed Hoyt, who owns this John Deere D, is ready to move on to other parts of the tractor.

Oil Pump Restoration

Any thorough engine overhaul should include an inspection of the oil filter body, oil pump, and oil supply lines.

A lack of oil pressure or low oil pressure can be caused by a distorted oil filter body, which, in turn, is usually caused by over-tightening the lower cover nut. The nut should be tightened only enough to eliminate oil leakage.

Like the gear pumps used in some hydraulic systems, the oil pump uses two gears that mesh together to create an area of high pressure on one side of the gears and low pressure on the other side. As one gear drives the other and the teeth mesh, oil is carried around the outside between the gear teeth and the housing.

Inspection involves cleaning or replacing the filter screen or element, depending on the model and configuration, then checking the gear surfaces and housing for wear and cracking. Be sure to also check for worn bushings or bearings and ensure that all oil passages are clear. Check your repair manual if you have any doubt about wear criteria.

Finally, make sure you use the correct thickness of gasket between the cover and pump body when you put everything back together. This not only prevents leaks but also provides end clearance for the gears. Ideally, the pump should also be primed before replacing the oil pan. One way to do this is to pack it with a lithium-based grease, commonly called "white lube," before replacing the oil pan.

Once you have the engine back together and running, you can finish by adjusting the oil pressure to the recommended level. On most horizontal two-cylinder models, it is between 10 and 15 psi, whereas it runs as high as 30 psi on Dubuque-built tractors.

Every thorough engine overhaul should include an inspection of the oil pump, since it provides the life blood to engine parts.

Pony Engines

The first use of a pony engine on a John Deere tractor was in 1949 with the introduction of the Model R diesel. Compared to gasoline and all-fuel engines, diesel engines require a lot more power just to turn them over, since it is the heat generated by compression that actually ignites the fuel. As a result, most manufacturers, including John Deere, equipped the tractors with a "pony engine"—a small, auxiliary engine that was started on gasoline. The clutch on that engine was then engaged with a drive on the flywheel to turn the diesel engine fast enough and long enough to get it started.

It wasn't until late in the 20 Series production that John Deere came out with a 24-volt electrical system and an electric starter large enough to do the same job. Even then, a pony engine continued to be an option through the 30 Series and the end of two-cylinder tractor production in 1960.

The first pony engine produced by John Deere, used only on the Model R, was an opposed, two-cylinder engine with bored cylinders. By the time the Model 70 diesel model came out in 1954, Deere had switched to an opposed, four-cylinder engine with sleeved cylinders. That engine was then used through the end of two-cylinder production.

According to Scott Carlson with the John Deere Collectors Center, most of the parts needed to overhaul both engines are still available, but they don't come cheap. Carlson says it's not unheard of for a restorer to spend as much or more overhauling the pony engine as he spends on the main diesel engine.

Missouri restorer Roy Ritter says the need for pony engine overhaul depends largely on how the engine was cared for by the original owner.

"I've seen a lot of guys start the pony engine, rev it up to full throttle, and engage the clutch before it even had time to warm up," he recalls. "When you put a cold engine under full load that many times, it's not going to stand up very well."

Fortunately, all of the procedures and specifications you'll need for pony engine restoration are included in most diesel tractor service manuals. Otherwise, the steps are similar to those outlined in this chapter for any gasoline engine. If you're lucky, you may be able to get by with a simple tune-up, which is what many restorers finally resort to doing anyway.

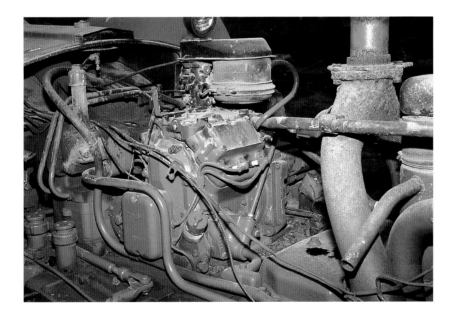

In many cases, engine tune-up or overhaul involves more than one engine. Until late in the production of the G and 20 Series, when 24-volt electric starters were introduced, diesel models were equipped with a two- or four-cylinder "pony" engine that is used to start the main engine.

CHAPTER 7

Clutch, Transmission, and PTO

You'll be comforted to know that transmissions were probably the toughest-built items on antique tractors. In fact, about the only time a tractor restorer faces a big problem is when a transmission failure was the reason the tractor was permanently parked in the tree row in the first place—and that's the bad news.

If the tractor is in running condition, you should have had the opportunity to drive it and run it through the gears before you made the purchase. Hence, you should have an idea where the trouble lies. Often it will be with one of the gears that saw a lot of field work, particularly if there were only three or four gears. Any clicking noise should be checked out, too, as it may indicate a missing gear tooth.

Other things to watch out for are the effects of dirt and water. Just as it is with a modern-day transmission, dirt is the gearbox's biggest enemy, which means that if dirt was allowed to get in the transmission, you may face some repairs.

Still, there's a good chance that you will get lucky and find that all it takes to get the transmission in working order is to drain it, clean it up, and refill it with fresh transmission fluid. In fact, most tractor enthusiasts say that's all that has been needed on the majority of the tractors they have restored.

A special cut-away tractor owned by Lester, Kenny, and Harland Layher shows the arrangement of the transmission and final drive gears. Note the cut-away cylinder between the bull gears that lifts the three-point hitch.

If the transmission appears to work satisfactorily, but is just a little noisy, replacing the transmission fluid with a heavier-weight product—going from GL5 to GL4 lubricant, for example—may take care of the problems. That's assuming you're only going to use the tractor on occasion for parades and shows. If you plan to use the tractor as a work machine, that may just cover a bigger malady.

The clutch is probably going to need more attention than the transmission. If during your test drive—assuming the tractor is in running condition—you found the clutch slipping or chattering, you'll need to take a closer look. Most likely the problem is a worn clutch plate.

Transmission Inspection and Repair

As stated earlier, transmissions used in tractors built in the 1920s, 1930s, and beyond were generally heavy-duty enough that they don't present any major problems, even after sitting for several years. You'll note, too, that John Deere didn't use helical gears in the two-cylinder tractors. All of the gears in the transmission and final drive system are straight-cut gears. Although straight-cut gears are not nearly as quiet or smooth shifting as helical gears, they do tend to be a little tougher, and they don't produce nearly as much side thrust. Still, it is a good idea to at least open up the transmission to check out the gears and clean them up.

When working on a horizontal two-cylinder tractor, you have the luxury of being able to overhaul the transmission and clutch without pulling the engine.

Several restorers say they like to drain any fluid they find in the transmission case and replace it with diesel fuel or kerosene. Keep in mind that after sitting for several years, the fluid can be about as thick as tar. While you're draining the transmission, keep an eye out for pieces of metal or fine shavings that can tip you off to problems.

If there are any teeth missing from a gear, it's important that you find them, too. As a John Deere mechanic for more than forty years, Missouri restorer Roy Ritter has seen what can occur if you don't locate the errant tooth. Everything may be fine on a warm day, he explains, but when the weather gets cooler and the fluid gets thick, it's possible for a gear to pick up the broken tooth with the oil. If the tooth ends up trying to go between two gears, you can bet the tooth isn't going to crumble. Something's got to give, and if it's not the shaft, it's the sidewall of the transmission case.

The next step is to drive the tractor around the yard for several minutes to circulate the diesel around the transmission case, coating and rinsing all the gears. If the tractor can't be driven, but can be towed, that would be your next-best option. As a final step, there's no substitute for using a stiff parts brush or paint brush and a can of kerosene or diesel fuel to clean the gears by hand, rotating each gear set as you go.

Don't forget to inspect and clean all the gears on the outside of the transmission housing. John Deere horizontal two-cylinder tractors are unique in that at least the first set of reduction gears, and in some cases other gear sets, are located under a cover installed on the cast housing. It's worth noting, too, that on A, B, and G tractors, the oil in the first reduction gear cover is engine oil that is shared with the crankcase. On the numbered tractors, starting with the 50, 60, and 70, the first-gear reduction case shares its oil with the transmission and final drive.

Next, inspect all the bearings, shafts, and seals for damage. Rotate the gears with your fingers as you go through the cleaning and inspection process. At the same time, check for looseness and rough action. If a gear wobbles or shakes, it's probably going to need repair.

Keep in mind that transmissions built in the early part of the twentieth century were not synchronized like they are today. So if farmers didn't wait until the tractor stopped before shifting, they had a tendency to round off the gears. This was particularly the case with the "road gear" in many late 1930s and 1940s tractors. Consequently, you may find a gear or two that needs replacement due to ground teeth. In the worst cases, you'll even find gears on which the teeth have been broken off due to "speed shifting."

While inspecting or rebuilding the transmission, it's also important to check for end play on the transmission shafts. On horizontal-engine tractors, there is generally no end play adjustment on the top shaft, since it rides on bearings. In most cases, the bearings are either fine as they are, or it's obvious they need to be replaced. However, the bottom shaft can be shimmed as needed on the right-hand bearing plate to remove any play.

The first steps in transmission overhaul are cleaning up the gears and shifting mechanism and inspecting the gears for missing teeth.

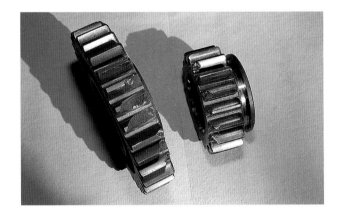

It's easy to spot the damage on these gears. The chipped teeth and rounded corners were caused by "speed shifting" the transmission before the tractor came to a stop.

Clutch Inspection and Rebuilding

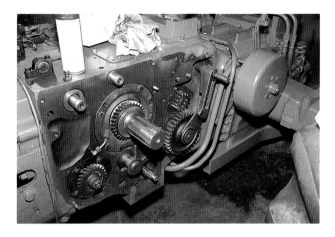

Whether the gear sets are on the inside or the outside of the transmission case (as is the case on later models with multiple gears), it's important to inspect the bearings and seals.

Be sure to replace the gasket before replacing the transmission housing cover.

Once you've finished with the transmission—either having cleaned it and verified its condition, or replaced the appropriate gears and bearings—it's time to move on to the clutch. Essentially, these two components work together.

Overhauling a clutch is important for more than accurate restoration. It's also a safety issue. Obviously, you want the tractor to stop when you push in on the foot clutch or pull back on the hand clutch and apply the brake.

While some of the early tractors, including steam tractors, had positive-engagement clutches, the move toward internal-combustion engines necessitated the use of a friction clutch that could be slipped as the drive was engaged.

One of the most notable features on a Waterloo-built John Deere tractor, besides the horizontal two-cylinder engine, is the external, twin-disc, hand-operated clutch. On the other hand, because they utilize a vertical-cylinder engine connected to a driveshaft, similar to other tractors built at the time, Dubuque- and Moline-built tractors incorporate a plate clutch that is attached to the engine flywheel. This means the tractor needs to be split at the bell housing or the engine removed for clutch replacement or inspection.

If you drove the tractor before you purchased it or started restoration, you should already have an idea whether the clutch needs attention. Usually the trouble is obvious and falls into one of three categories:

• The clutch slips, chatters, or grabs when engaged
• It spins or drags when disengaged
• You experience clutch noises or clutch handle/pedal pulsations

If the clutch slips, chatters, or grabs when engaged, or drags when disengaged, the first thing you should do before tearing into the clutch itself is see if the clutch linkage is improperly adjusted. In some cases, free travel of the clutch lever is the only adjustment necessary for proper operation of the clutch.

On horizontal-engine tractors with a hand

Above: On Dubuque-built two-cylinder tractors, it's necessary to split the tractor between the engine and transmission to repair the clutch, just as you would with any vertical-cylinder tractor.

Right: The transmission on the low-horsepower Dubuque-built tractors didn't feature many gears. Hence, it could be positioned right in front of the differential.

clutch, there should be a definite feel of over-center action when pushing forward on the hand lever. A slight pressure should be felt in the hand lever, then a definite release of pressure as the clutch engages.

Foot-Clutch Systems

Inspection

Beyond clutch-linkage adjustment, the cause of most problems on vertical-cylinder tractors equipped with a foot clutch is found in the clutch housing, where clutch components are either worn, damaged, or soaked with oil that is causing the clutch facing to slip. Most often, the culprit is weak or broken pressure springs, or worn friction disc facings. Hence, clutch restoration generally consists of cleaning and checking all parts for wear; replacing bushings in the sleeve, if necessary; and replacing the clutch lining.

Clutch Throw-Out Bearing

Another potential problem spot on vertical-cylinder tractors is the throw-out bearing, which compresses the springs to release pressure on the plates when the foot clutch is pushed in. Characteristics associated with a defective throw-out bearing include rough actuation and squealing when the clutch is released. Replace the throw-out bearing if there is any hint of roughness, looseness, or discoloration.

The manual for the Model M and MT also notes that early production tractors were equipped with a bushing that served as the pilot bearing for the forward end of the clutch shaft. On later-production tractors, the bushing was replaced by a ball-type pilot bearing. If it becomes necessary to replace the bearing, you can convert the early production bushing-type to the ball-type by installing the late-model clutch shaft, pilot bearing, grease retainer, and pilot bearing adapter.

Once you have everything back together, the clutch-pedal free travel can be adjusted by rotating a cam, found on the left side of the center frame, in a clockwise direction. On most Dubuque-built models, clutch-pedal free travel of 1½ inches is recommended.

Hand-Clutch Systems

Inspection

Restoration of the hand-clutch system found on horizontal-cylinder tractors consists of inspecting all clutch and fork bearings and replacing any clutch facings that are worn, badly glazed, or oil soaked. You should also replace any facing that bends easily. A facing that is in good condition should be rigid.

However, the first step is removing the clutch-drive components, including the drive disc. To do this, first remove the cover and clutch adjusting disc by removing the three adjusting nuts referred to in clutch adjustment. Next, remove the clutch release springs and the linings, discs, and facings. Some models, like the early As and Bs, will have only a facing disc between the adjusting disc and the drive disc; others, like the D and late-model A, B, and G, have a facing disc, as well as a sliding drive disc and a lined disc between the adjusting disc and drive disc. Virtually all models have a second, inner disc between the drive disc and the belt pulley.

Next, you'll need to remove the driving disc. But first, check to see if the clutch-drive disc is stamped with the letter V that matches up with another V on the crankshaft to form an X. If the V's are missing, make the appropriate correlation marks before removing the disc. On the Model D, you'll also need to loosen the drive disc clamp screw.

On some models, you can use two ½-inch bolts as a puller to remove the clutch drive disc. Just be sure you rotate the pulley so that the ends of the bolts do not rest on the swedged head of the clutch cone drive pin when the bolts are tightened. On other models, you can use two cap screws and a handmade plate or some other suitable puller to remove the disc.

Before replacing any other parts, you'll need to make sure the springs meet the specifications listed in your repair manual and ensure that springs are not rusted or distorted. In many cases, new fork bearings and new clutch facing plates are all that are needed to restore the two-cylinder clutch to operating condition.

John Deere recommends replacing any clutch facings that are worn, badly glazed, or oil soaked.

Adjustment

In practice, the hand clutch is engaged and disengaged by a set of clutch dogs that engages or releases the spring pressure on the facings and clutch plates. At the same time, the clutch lever actuates a drum brake each time the clutch is disengaged.

Like any clutch, the hand clutch can slip and cause a loss of power, overheating, and damage to the clutch facings if it is not adjusted correctly. Fortunately, this adjustment is a relatively easy process on most Johnny Poppers. Start by removing the belt pulley dust cover, as well as the cotter pins from each of three clutch-adjustment bolts. Then, with the clutch in the engaged position, tighten the nuts evenly a little at a time and to the same tension. Check the tightness of the clutch after each adjustment by disengaging and re-engaging the clutch. On most models, the clutch lever will produce a distinct snap when engaged if the adjustment is correct.

In some cases, such as on early Model G tractors, there is no snap when the clutch goes into engagement. In these cases, you'll need to measure the amount of pressure required at the end of the control lever to lock the clutch in the engaged position, then adjust accordingly. Your tractor service manual should tell you how many pounds of pressure it takes to lock the clutch in the engaged position.

The John Deere clutch is engaged and disengaged by a set of clutch dogs that engage or release the spring pressure on the facings and clutch plates.

The broken clutch dog on this clutch/belt pulley will need to be replaced as part of the restoration.

Pulley Brake Adjustment

Since the pulley brake is controlled with the same lever as the clutch, it will be necessary to separately adjust the brake pulley once the clutch setting has been established. To do this, simply loosen the adjusting-screw lock nut and turn the adjustment screw to the point at which the pulley brake will stop the belt pulley when the clutch is disengaged and the clutch lever is held back. Then, retighten the lock nut.

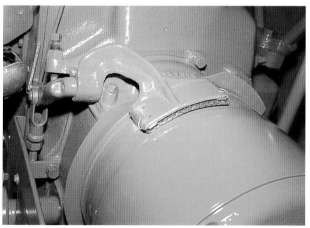

Power Takeoff Repair

Although the first rear power takeoff (PTO) appeared as early as 1918, the PTO didn't play a big role until a few decades later. Until rear-mounted PTO-driven implements began appearing in earnest, most farmers continued to use the belt pulley as their main power source.

Early PTO systems were inconvenient, at the least, to use with certain implements, since the systems were driven from the transmission. That meant the tractor had to be moving or in neutral with the clutch engaged for the PTO to operate. Many farmers can still remember mowing hay fields with a transmission-driven PTO. If the sicklebar started to plug, or you had to stop in the middle of the field, you could almost count on getting it plugged even worse, because the minute you stopped, so did the sickle.

Things changed when Cockshutt introduced the first commercially available tractor with a live PTO in 1947. The live, or independent, PTO utilized its own clutch within the transmission, which meant the PTO could continue to operate in relation to the engine speed rather than slow or stop its function as the tractor slowed or stopped. Within a few years, all tractor makers had a live PTO, including John Deere, which introduced the feature shortly after the debut of the Model 50, 60, and 70 tractors.

Whether your tractor is equipped with a transmission-driven PTO or an independent-drive system, you should check its condition as part of transmission inspection and repair. The most common problem tends to be seal leakage. Other ailments can include clutch problems with live PTO systems and worn gears and bearings. Due to the variations used on the many different models, overhaul procedures are best explained in your tractor repair manual.

Top left: **John Deere horizontal two-cylinder tractors incorporate the clutch with the belt pulley. The clutch is adjusted by evenly tightening each of the three nuts on the adjusting disc.**

Bottom left: **It's important to check and adjust the pulley brake—replacing the brake pad if necessary—as part of a horizontal engine clutch restoration.**

CHAPTER 8

Final Drive and Brakes

In some respects, it would seem that brakes and final drives are two different subjects, and, therefore, shouldn't go together in one chapter. But when you consider that the role of the brakes on a tractor is to stop one or both rear wheels, you can understand how the two fit together.

In general, there are two points of application with the brakes used on most John Deere two-cylinder tractors. What's more, those braking points have a lot to do with the type of final drive you find on the tractor.

Most Waterloo-built models that used bull gears to drive the axles have brake housings located on the sides of the final drive case. Generally, these are expanding-drum brakes on a splined shaft that engage the bull gears. So, in effect, the brakes aren't stopping the axles or the wheels; instead they are stopping the bull gears that drive the axles. The exceptions are the Model H and certain high-clearance models, including the AH and GH. The H utilizes drum brakes splined to the axles, while the AH and GH use drum brakes that engage the drive chains.

In contrast, tractors built in Moline and Dubuque, which utilize a pinion shaft and ring gear in combination with a spider gear set, don't have a braking point on the gear sets themselves. Hence, the brakes are generally located on the drive axles or in the wheel hubs. Instead of drum brakes, virtually all Dubuque models use disc-style brakes splined to the axles.

While most vertical two-cylinder designs employ brakes on the axles, the majority of Waterloo-built tractors utilized externally mounted brake housings that applied braking pressure to the final drive bull gears.

Differentials and Final Drives

Although John Deere used several different types of final drives, you can count on all of them having one thing in common: They were generally built tough enough to take all the torque the engine and transmission could generate, and then some. As a result, the final drives on most John Deere two-cylinder tractors need little attention other than replacing bearings and seals and changing the fluid.

With the exception of the Model H, most Waterloo-built two-cylinder models employ a differential connected to a pair of bull gears in the final drive housing to drive the rear axles. Since the transmission is already transversely mounted, the main function is to divide the power being delivered to the axles and get both axles turning in the same direction.

In contrast, the final drive on vertical-cylinder, Dubuque-built tractors more closely resembled that of other tractor brands. In other words, it incorporated a hypoid- or bevel-gear-type differential to transfer engine torque from the transmission to the axles. Composed of a bevel pinion and shaft, bevel ring gear, and a set of pinion and side gears, the differential provides a means of turning the power flow 90 degrees and dividing the power between the two rear wheels. The differential also provides further gear reduction, beyond the choices provided by the transmission, for additional torque to the rear wheels.

Dubuque-built tractors and high-crop A and B models utilize a set of bull gears at the wheel end of the axle, rather than the final drive housing. In addition to providing gear reduction, the final drive gear set and housing serve to raise the tractor for additional ground clearance.

You'll find that some tractors still use chains as the final drive. Most notable are the GP and early D models. While the chains aren't likely to need replacement, due to their rugged construction, they may need adjustment. You'll notice, though, that John Deere didn't use any type of idler. Instead, the axle housing incorporates a concentric casting, which is attached to the rear end/transmission housing by a series of mounting clamps. To tighten the chains, loosen the clamps from the inside of the housing and rotate the axle housing. The service manual for the D recommends one inch of chain slack at the tightest position.

Because the bull gears on Waterloo-built tractors are extremely durable, most need only a good cleaning or, at the most, new bearings to restore them to like-new condition.

A number of Dubuque-built tractors, including this Model L, utilize an enclosed gear set at the wheel end of the axle to provide further gear reduction, while increasing ground clearance.

Some of the earliest John Deere tractors, including the D and GP models, used heavy chains to transmit power from the transmission to the axles.

Eccentric-shaped housings, held in place by clamping brackets around the circumference, allow the drive chains on this Model D to be tightened or loosened.

Differential and Final Drive Inspection and Repair

Inspecting and rebuilding the differential and final drive are not too much different than working on the transmission. Basically, the process means draining the old fluid, if it is different than that in the transmission; cleaning the gears; and checking for worn bearings and seals and missing gear teeth. Because of the vast difference between Waterloo- and Dubuque-built tractors, you'll need to refer to your tractor service manual for adjustment procedures and end play or backlash tolerances.

In most cases, the gear mesh and backlash of the main drive bevel gears are controlled by shims on the shaft. Generally, tooth contact and backlash should be checked and adjusted, if necessary, whenever the transmission is overhauled; it is imperative, however, when a new pinion or ring gear, or both, are installed. Keep in mind that neither the differential spider/bevel pinion nor spur ring gear is available separately on most models. If either part is damaged, you'll have to replace the entire assembly.

If shim adjustment is in order, the first step is to arrange the shims to provide the desired backlash between the main drive bevel pinion and ring gear as specified in your tractor service manual. The next step is to adjust the shims to provide proper tooth contact or mesh pattern of the bevel gears. Once you've obtained the ideal tooth contact, you'll need to re-check the backlash obtained in the first step to make sure it is still within the specifications.

One tip offered by restorers is that if you find bull gears with excessive wear, you can often swap them side for side. In effect, you're positioning the gears so that what was the gear lash in the forward direction is now the gear lash for reverse and vice versa. To remove the bull gear on most Waterloo-built models like the A, B, or G, first loosen the nut on the inner end of the axle shaft. Then, drive a long, tapered wedge between the inner ends of the axle shafts. Remove the nut and the axle shaft.

You'll need to remove the axles and bull gears to reach the spider/bevel pinion and spur ring gear on Waterloo-built models. Each of these parts is visible just ahead of the bull gears.

Axle Shafts

Like all other components in the final drive, the rear axles on early farm tractors were built tough. That doesn't mean they won't need attention, though.

One of the most common problems restorers encounter is an oil leak where the axle exits the rear axle carrier. This is most commonly caused by a failure of the inside seal. With few exceptions, most tractors use at least two seals in the axle housing. However, the outside seal is usually a dust seal, which will not stop oil leakage from the inside.

Of course, if the seal failed, there's a chance that the bearing needs to be replaced, as well. To check the bearings, raise the axle, and, with the wheel off the axle and the transmission in neutral, check for excessive axle shaft end play by attempting to move the axle in and out and up and down. On several models, including the 20 Series, it will be necessary to use a hammer and cold chisel to remove the outer felt seal retainer from the axle shaft.

Roy Ritter, a knowledgeable Deere enthusiast from Savannah, Missouri, notes that when you replace an inner seal, you need to be aware that there is a sump molded into the axle housing on most models. In the majority of cases, this cavity fills with oil during a leak, at which point the oil starts seeping out of the dust seal on the end of the axle.

Most people don't even know the sump is there, he says. But if it's not cleaned out during seal replacement, it can lead to confusion later. What often happens, he explains, is the tractor is eventually operated on a slope, which causes the oil in the sump to run to the end of the axle and leak past the dust seal. As a result, the restorer thinks the seal is leaking again, when, in reality, the seal is fine. You just missed the oil that was hidden in the axle cavity.

Another potential problem is that the defective seal may have worn a groove into the axle. Not only will this groove contribute to the leak, but it will also make it impossible to repair with a new seal. Hence, it's important to check the axle, just as you would any shaft on which a seal is being replaced. If you can feel the groove with your fingernail, it needs to be repaired.

While some shafts can be repaired with a Speedi-Sleeve, which is basically a thin collar that slips over the original shaft to create a new surface, a drive axle will generally require the second option. That is filling the groove with a metal-filled epoxy, such as J-B Weld. Two or three thin coats are usually better than one heavy one. Once the epoxy has cured, simply sand it to a smooth and symmetrical surface, finishing off with extra-fine sandpaper or emery cloth.

Even if the seals aren't bad, it's a good idea to replace them as part of the restoration process. It will be a lot easier and less expensive to do it now than to have to come back and do it later.

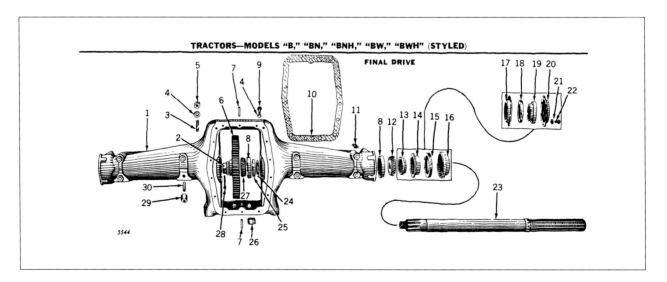

This exploded drawing of a Model B rear housing and axle gives you an idea what to expect when replacing the various seals and retainers.

Final Drive and Brakes / 107

The grease built up on the inside of the drive wheel is a clear sign that this tractor will require new drive axle seals.

With few exceptions, most tractors use at least two seals at the end of the axle housing.

If damaged or worn seals have cut a groove in the axle shaft, you'll need to pull the axle and repair it with a sleeve or durable epoxy.

With new seals installed on the ends of the axle, this classic two-cylinder is ready for the wheels.

Brake Restoration

The brake system is not the place to cut corners. You might be able to get by without overhauling the transmission or opening up the rear end, but you'll want to give the brakes the attention they deserve. This isn't just for your safety, but also for the safety of those around you. If you take your tractor to any shows at all, you're going to need to unload it off a truck or trailer. And this is not the place to have your brakes fail to hold. Consider, too, how you would feel if you lost control of your tractor in a parade, with people lining both sides of the street. Fortunately, brake restoration is not a difficult job.

Interestingly, a number of early John Deere tractors, including the GP, early D models, and BR models built prior to 1938 were equipped with only a single brake. This was primarily used for parking or holding the tractor in place while it was being used as a belt pulley power source. However, with the introduction of row-crop tractors like the A and B, individual brakes on the rear wheels became standard equipment.

If you haven't already noticed, the brake drum cover on early model Waterloo tractors was riveted onto the unit. In the late 1940s, the cover was redesigned so that it could be more easily serviced. As a result, it is necessary to remove the entire "riveted" brake unit from the early tractors, while later models with a bolt-attached cover can be inspected and repaired without removing the unit from the tractor.

Before you begin, remember that brake springs can fly off unexpectedly and in who-knows-what direction if they are not handled properly. Always wear safety glasses when attempting any work on the brakes. Since you will no doubt be lifting the rear of the tractor off the ground, you should also double-check to make sure it is secure from rolling or tipping.

While it's important that you start brake troubleshooting by checking all brake adjustment points, it's also important to examine the brake linings or pads. If brake linings aren't badly worn or oil soaked, you may be able to bring them back to life by simply roughing them up with sandpaper. Other restorers

If a brake kit is no longer available for your tractor model, any automotive shop that does brake work should be able to rivet new linings in place on your brake shoes.

If you're lucky, you may be able to bring the brakes back to life by simply roughing up the brake shoes with sandpaper or using a torch to dry out oil-soaked shoes. In this case, though, the job called for new shoes and a thorough clean-up.

have used a torch to dry out oil-soaked brake shoes that have sufficient wear left on them. You should also make sure the brake drum or brake disc is clean and free of rust. But again, if it isn't, a piece of emery cloth or sandpaper applied to the problem area will fill the bill.

If you do find that the brake shoes or linings have become oil soaked or worn too far, it should be fairly easy to find replacement pads, linings, or shoes. If a brake kit is no longer available, any automotive shop that does brake work should be able to rivet new linings in place on your brake shoes. The same shop should also be able to turn any problem brake drums on a lathe, removing any grooves or out-of-round spots.

Above: If the brake engages the bull gears on the final drive, or extends into any other oil-enclosed housing, make sure you install a new gasket during assembly.

Right: After the brakes have been refurbished and reinstalled, adjust the brake to the point at which the rear wheel begins to drag, then back off slightly.

Brake Adjustment

Regardless of whether or not the brake pads or discs have been renewed, touched up, or approved in the current condition, it is important that the brakes be adjusted. Moreover, foot brakes that are adjacent to each other, or can be locked together, such as those found on the M and MT, need to be adjusted to a comparable setting with an equal amount of free travel. Otherwise, one brake may be applied while the other just drags. This isn't a problem, of course, with hand brakes or on tractors where the left and right brake are operated by the corresponding feet, as on the A and B.

On most horizontal-engine models, on which external brakes contact the bull gears, John Deere advises tightening the adjusting screw on the brake drum housing until you obtain a free travel distance of approximately 3 inches. The same holds true for the Model H, even though the brake assembly is enclosed in the rear axle housing.

Most Dubuque-built tractors utilize disc-type brakes that are splined to the outer end of the final drive shafts. The brake-adjustment procedure varies with the model, but in general, it involves turning the adjustment rod or the adjustment nut until a slight drag is obtained when turning the wheel, which has been raised off the ground. The rod or nut is then backed off as indicated. On the MT, the service manual calls for free travel for the first 1½ inches, with wheel lock occurring at approximately 2½ inches. Meanwhile, the brake pedals should be synchronized by varying the length of the individual pedal rods.

CHAPTER 9

Front Axle and Steering

Steering Through Deere History

Few tractor manufacturers offered as many front axle options and configurations through history as John Deere did in the early 1900s. As was the case with nearly all tractor models, the first John Deere models were equipped with wide front axles only. This was basically a carryover from the steam-tractor era in which tractors were used for plowing and belt applications. Even at the time the Model D was introduced, tractors were still being used for wheatland-type applications, like pulling a plow, disc, or grain binder, and for rugged jobs like pulling stumps. Cultivation was still done by a team of horses. Of course, belt applications were still very common, which fit the D just fine.

By the 1930s, most manufacturers, including John Deere, had introduced a tricycle configuration as standard equipment on certain models and as an option on others. The trend started in 1924, when International Harvester introduced an "all-purpose," or tricycle-type, front end on the McCormick-Deering Farmall. Deere quickly followed with the GP, which offered the same clearance, but not the narrow front end. After trying to remedy the situation with the GP Wide Tread, Deere finally hit a home run with the A and B, which featured a narrow front end as standard equipment. By the time World War II ended and many veterans were going back to the farm, some dealers had a hard time selling a tractor with a wide front end.

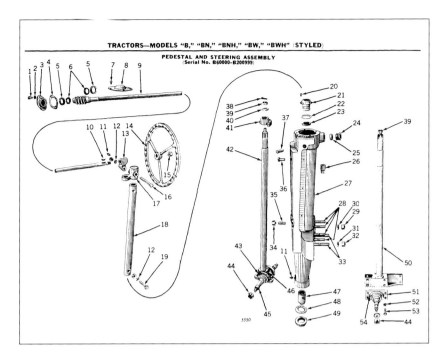

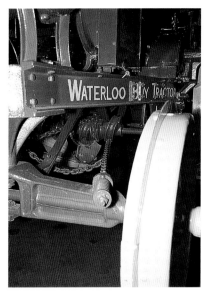

Illustrated above is the typical pedestal and steering assembly used on a narrow-front, letter series tractor.

Like the steam engines that preceded it, the first Waterloo Boy employed a set of chains wound around a steering shaft to turn the front axle.

Later-model Waterloo Boy tractors, as well as the first Model D tractors, used a gear on the end of the steering shaft to rotate a geared arm on the front axle.

Because of their different design, Dubuque-built, vertical two-cylinder tractors utilize a gear housing at the end of the steering shaft to turn the front wheels.

One of the steps in steering system restoration is checking for free play in the bolster on narrow-front models. Desired end play on this John Deere model is controlled by shims under the gear.

Not only did the tricycle front end fit between corn and vegetable rows of the time and make shorter turns at the ends of the field, but most farmers also liked the visibility the narrow front axle offered when using a front-mounted cultivator. Like other manufacturers, John Deere used its tricycle design as a base for other implements, like a two-row corn picker that mounted directly on the tractor.

Unfortunately, tractors with a narrow front axle also proved to be more dangerous than those with wide axles, especially if they were top heavy or turned too fast on sloping terrain. Times were changing, too. As equipment got bigger and the turning radius improved on standard tractors, farmers found they didn't have to turn such tight corners. They also discovered that they could guide the tires on a wide front axle between the rows as easily as they could keep dual tricycle wheels in a single row width. Besides, rows were getting narrower.

Steering Configurations

Mention steering in connection with a John Deere two-cylinder tractor and most people naturally think of the long steering shaft routed across the top of the hood, or through the hood and above the engine. The ultimate picture of simplicity, the shaft had a steering wheel on one end and spines on the other end, which connected to a gear head at the top of the steering pedestal. That didn't mean John Deere didn't use other types of steering mechanisms or configurations, though.

If you go back to the early Waterloo Boy tractors, you'll find a steering shaft that is wound with a pair of chains connected to the front axle. As the steering shaft is turned, the shaft rotates to wind up the chain on one side while providing slack on the opposite side. Later, both the Waterloo Boy and the first Model D tractors were equipped with a steering shaft that directly engaged a geared steering segment on the axle with a worm gear on the steering shaft. Turning the steering wheel took some force. It wasn't something you did while the tractor was sitting still.

Through most of the two-cylinder production era, though, John Deere tractors were equipped with one of two types of steering mechanisms. The tractor either utilized a pedestal with a steering sector at the top of an enclosed steering spindle; or it employed a steering gear box that connected to a vertical steering shaft, which, in turn, connected to a steering arm. The former is found on virtually all narrow-front tractors, as well as wide-front (AW, BW, AWH, BWH, GW, and HWH) models. The latter is found on later Model D tractors as well as specialized versions such as the AR, AI, and BR.

Regardless of the configuration, restoration will primarily consist of replacing worn parts, bearings, bushings, and seals, and making the appropriate adjustments. This will include adjusting for wormshaft end play; vertical shaft or vertical spindle end play, depending upon the model; and backlash. In each case, it will be necessary to support the front of the tractor to remove the load from the steering gear.

It's important to note that with some models, such as the D, wormshaft end play and vertical shaft end play can be adjusted by simply loosening a lock bolt and tightening the adjusting nut. Pedestal models, such as the A and B, however, require the addition of shims to make the appropriate adjustments. Your service manual will provide the proper technique for overhaul, as well as the specifications for adjustment.

Front Axle Repair

The life a vintage tractor led before you acquired it has a lot to do with its condition and the repairs that it is going to need. The condition of the steering gear and front axle is a textbook example. A tractor that spent most of its days in large wheat fields in western Kansas, for example, isn't going to have near the wear on the knuckle bushings and axle pivot pin as a row-crop model that spent every working hour crossing corn furrows and turning around on end rows. This is just one more reason to find out all you can about the tractor before you make a purchase or calculate repair costs.

That said, the only satisfactory way to overhaul the front axle assembly on a wide-front tractor is to remove it from the tractor and make a complete check of all bearings and bushings. That includes the center pivot pin, steering arms, ball seats on tie rods, knuckles, and wheels. It's important to note that on some tractors, the knuckle post bushings are pre-sized to provide a specified amount of clearance for the knuckle post.

On models with a tricycle-type front end, inspection primarily consists of checking and replacing wheel bearings and seals. In fact, on most models, it's possible to remove the axle assembly or bolster assembly from the bolster, making repair or replacement of the axle even easier. Depending upon the amount of free play in the bolster, it may also be necessary to replace bushings, bearings, and seals in the upper or lower sections of the bolster.

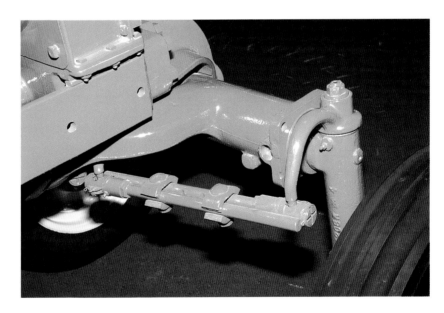

Be sure the drag links are tightly secured and adjusted for proper front wheel toe-in.

Quite often, the kingpin or pivot shaft, on which the wide front axle member pivots, is the first thing that wears out or needs replacement.

Before installing new bearings, seals, and felt washers on the axles, make sure the spindle is clean and free of nicks and scoring. A piece of fine emery cloth can be used to remove any rough spots or rust.

In most cases, the bolster and horizontal axle are available as individual units, which makes it easier and less expensive to replace a damaged axle.

Roll-O-Matic Axle

In 1947, John Deere introduced an innovative new option on tractors equipped with a tricycle-type front end. Called the Roll-O-Matic, this unique narrow-front assembly was intended to help smooth out rough fields by moving the front wheels in opposite directions when a bump or ditch was encountered. This was done by gearing the two wheels together in such a manner that when one wheel went up, the other wheel was forced down an equal distance.

In general, overhaul of the Roll-O-Matic system is not much different than the conventional knuckle-mounted dual front wheels except that there is an extra bushing, retainer, and seal where the Roll-O-Matic knuckle attached to the pedestal extension. It's important that the knuckle units on the Roll-O-Matic assembly are installed so the timing marks on the gears are in register. The center housing should also be packed with wheel bearing grease before the final knuckle is reinstalled.

Power Steering

On all Waterloo-built tractors, the power steering is primarily incorporated into the steering components. The valve, for example, is located above the steering pedestal, where it is actuated by the steering shaft itself. The steering cylinders, meanwhile, are built into the pedestal. Finally, the hydraulic pump is mounted on the fan drive shaft, which means there is minimal travel of the hydraulic fluid between the pump, valve, and cylinders. As with anything, though, there are exceptions, which should be confirmed via a good service manual that is applicable to your tractor. The 620 orchard tractor, for example, uses a power steering cylinder and valve that receives its working fluid from the Powr-Trol pump.

As with the hydraulic system, the maintenance and repair of the power steering system requires absolute cleanliness of all parts. It's also important that all parts be free of nicks, burrs, or scoring that can affect performance. To inspect and service the system, it's best to follow the procedure outlined in the tractor service manual, since service involves checking pressures; adjusting flows and operating pressures; and inspecting various valves, washers, springs, and end play clearances.

Both photos: Offered as an option on some models, John Deere's Roll-O-Matic tricycle front axle helped smooth out rough fields by moving the front wheels in opposite directions when a bump or ditch was encountered.

Waterloo-built two-cylinder tractors have all the components of the power steering system, except for the pump, incorporated into the steering pedestal.

Rather than using a cylinder to turn the axle, Waterloo-built tractors use a unique system that consists of two steering vanes within a cylinder at the base of the pedestal.

Power steering pump maintenance should be handled in the same manner as hydraulic system inspection and repair.

Cleanliness is next to godliness when it comes to restoring steering components.

The scoring on the inner surface of this power steering gear pump body calls for a replacement of both the body and the gears.

It took a while to figure out why the power steering wasn't working properly on this 20 Series tractor. The problem was a broken pin on the steering vane.

Turning the flow control valve adjusting screw inward will cause faster, easier steering action. However, this can also cause a decrease in front wheel stability, which contributes to front wheel fluttering. Front wheel flutter can also be caused by improper adjustment of the steering worm and sleeve assembly.

Steering Wheel Repair

Unless your tractor has been protected from the elements for most of its life or you have a GP or an early D with a steel steering wheel, there's a pretty good chance the steering wheel is going to be severely cracked. Fortunately, there are several solutions available to you. Should you prefer to have the steering wheel professionally repaired, there are several companies that specialize in refurbishing steering wheels. Minn-Kota is one of the most notable; it can take your old wheel and mold new plastic around the steel rim, complete with the original grooves, ribs, or finger ridges.

Should you choose to repair a cracked steering wheel yourself, there are a couple of options practiced by restorers. One professional restorer says he uses Fiberstrand body filler to fill all the cracks and crevices. Another uses a body filler such as Evercoat polyester glazing material and follows that with a coat of fast-fill primer. With either product, the steering wheel must be sanded smooth after the material hardens and then painted.

You can always buy a reproduction or refurbished steering wheel as well. Just be sure you get the right style for your particular model, assuming authenticity is important to you. Most unstyled models featured flat-spoked steering wheels, while styled tractors featured round spokes.

For a professional look, consider sending the steering wheel to one of several companies that specialize in refurbishing plastic or wood-rimmed steering wheels.

CHAPTER 10

Tires, Rims, and Wheels

Let's be honest. Tires and wheels can be a costly and frustrating part of a restoration project. If your vintage John Deere tractor was originally equipped with steel wheels, there's a good chance that rust has taken a toll. There's also a chance that a previous owner or backyard mechanic removed the steel rims so the tractor could be fitted with rubber tires. Although some "shade-tree mechanics" went to the trouble of cutting the steel rims off the spoked wheels before welding the spokes to drop rims designed for rubber tires, others simply ordered new wheels through their John Deere dealer.

According to historians with the Two Cylinder Club, the specifications for standard-equipped D tractors, for example, were changed from steel wheels to rubber-tired wheels in November 1940—even though nearly 52 percent of the tractors sold since July 1, 1939, were equipped with all-rubber tires. After 1940, steel wheels for many John Deere two-cylinder tractors continued to be available with a variety of extension rims, lugs, and grousers as special equipment. Hence, you may need a little help finding out what was original, what was optional, and what was "engineered in the field."

In the meantime, you'll need to decide whether you want to keep the rubber tires or go back to steel wheels to attain the original look of the tractor. That's not the last of your decisions, though. If the tractor was equipped with rubber tires from the factory, chances are the tires have been replaced with different tires than originally came on that model. If you're planning to use the restored tractor as a work tractor, that may not matter to you. But if you plan to restore it as a show tractor, you may want to find the correct components.

In an effort to convert steel-wheeled tractors to rubber tires, shade-tree mechanics often welded tire rims to the old spokes and added a few extra braces.

Rear Wheel Removal

Most row-crop John Deere two-cylinder tractors are equipped with an adjustable rear axle that allows the wheels to be moved in or out for variable tread widths. If you're lucky and the tractor has been well cared for, you should be able to slide the rear wheels on the axles, or even off the axles, in the intended manner.

First, loosen all the hub bolts, removing two of them altogether. Now, install the two bolts you've removed in the empty holes on the hub. These will serve as "pullers." Finally, jack the tractor up enough to raise the wheel off the ground and begin tightening the puller bolts. Ideally, you should be able to tighten the bolts sufficiently to break the hub loose.

Unfortunately, wheel hubs are like a lot of components on classic tractors. If they have sat long enough without being moved, there's a good chance they have solidly rusted into place. If this is the case, you'll need to start the removal process by saturating them on a regular basis with some type of penetrating oil. If after trying to remove the hub in the traditional manner, it still doesn't budge, you may need to heat the hub with a torch to expand it. Don't expect a propane torch to do it, though. The hub will absorb a lot of heat, so an acetylene torch is usually more appropriate. Don't beat on the hub when it's hot. This will only deform the parts and make matters worse.

You may have to fabricate a puller that will butt up against the end of the axle. The best way to do this, according to Gene Tencza, a Deere enthusiast from Orange, Massachusetts, is to start with a piece of steel approximately an inch thick and drill holes in it to match the hub bolt pattern. Then find some long bolts to extend through the holes to the hub. If the hubs are very far from the ends of the axle, you may need to substitute pieces of threaded rod on which you have welded a nut. It will take a lot of bolt turning to pull the wheel off, Tencza explains, and you'll have to stop when the wheel gets to the end and add an extension on the axle to finish the job.

Once the wheel and hub have been removed, by whatever means necessary, you can clean and smooth the axle so that the wheel will slide effortlessly when reinstalled.

Although John Deere incorporated a "puller hub" into the adjustable rear axle, the wheel can still be difficult to remove after years of accumulated rust.

Tire Repair and Restoration

If the tractor has been sitting for some time, it's quite possible that the rims have been corroded by calcium chloride. Or perhaps the tires have rotted away after years of sitting in a pasture and the wheels came into contact with the ground, allowing rust to take its toll.

If you need one more thing to worry about, it's the fact that replacement tires can be rather expensive, especially if you're restoring a behemoth like the Model R, Model 80, or Model 820. All three featured rear tires that were either 14-35 or 15-34, with a 18-26 tire available as an option. Although the 15-34 size is no longer available and would have to be replaced with a 16-9x34 tire, any new set of tires that fits the old rims will set you back at least $800.

If you're only looking for a working tractor, any tires that fit the rim will generally be acceptable. The newer-style tires, with their 23 degree bar, or long-bar/short-bar design, may even offer better traction than the 45 degree lug tires originally found on the tractor.

However, if your goal is to restore a vintage model to show condition, and you're after accuracy, the challenge is a little greater. Not only did the tire companies change their size standards, but they also changed tire styles as more effective patterns were discovered. Fortunately, there are several independent sources for the most sought-after sizes and types of tractor tires. They include M. E. Miller, Gempler's, and Wallace W. Wade. (See the list of parts sources in the appendix for addresses and phone numbers.)

One option for repairing tires that are worn or slightly damaged, but still usable, is to install a set of tire reliners. These are made from old tires that have had the lugs ground down, making the reliner itself about ¼-inch thick. Generally, reliners come in half-moon shapes so they can be easily inserted into the old tire, where they partially overlap. Some companies also offer spot reliners. The inner tube is inserted into the tire, where, once inflated, it holds the reliners in place.

Other tire repair products include rubber putty that can be used to repair cracks, gouges, weather checking, and other minor problems. Unfortunately, it can't be used to repair a hole.

Finally, various companies carry a concentrated black tire paint that can be used to revive the color of old, gray-looking tires. Simply mix it with paint thinner according to the directions on the label and apply it as you would paint.

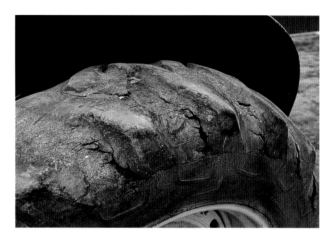

Even though this weather-cracked tire still holds air, it's not very pretty to look at. Tires are a costly investment, but it makes sense to replace the pair as part of the restoration.

If you're restoring a work tractor, consider using newer-style tires. With their 23 degree bar, they generally offer better traction than the 45 degree lug tires originally found on the tractor.

Wheel and Rim Restoration

As you'll quickly discover, John Deere used a variety of wheel sizes and types on their two-cylinder tractors over the years, which can lead to a lot of confusion about what wheels should be considered original. Many early steel-wheel models used what are known as "skeleton wheels." Basically, these are very narrow wheels that have cast lugs bolted to the rims in an alternating offset pattern. For additional flotation, Deere also offered a wider rim with lugs that bolted directly to the flat part of the rim. There was even a rim extension available, which could be attached to the standard rim with a set of clips to make the wheel even wider.

By the mid 1930s, Deere was offering rubber tires on most models, yet the wheels still featured spokes similar to those used on the steel wheels. It wasn't until 1936 that Deere made pressed steel wheels available, and spoke wheels were still an option for a few more years. However, that option changed in late 1939, when Model A tractors ordered with spoke wheels for rubber tires got flat spokes. Prior to that, the spokes had been round. The final step in the evolution of rear wheels was, of course, the introduction of cast rear wheels to which a demountable rim was attached with a set of rim clamps.

Front wheels, meanwhile, went through a similar evolution. Early options for many row-crop tractors included cast disk wheels, as well as spoked steel wheels that could be fitted with both guide bands and extensions. The former not only helped the front end track through turns, but, with the aid of even higher guide rings, also helped the operator hold the tractor on the top of listed corn rows.

However, by 1937, round-spoke front wheels were discontinued and replaced with pressed steel wheels and rubber tires on most models. As a result, you'll see archive photos of John Deere tractors built in the late 1930s with spoke front wheels and pressed steel rear wheels and vice versa.

If you've been fortunate enough to find rims and wheels that are in relatively good shape, the restoration process may be as simple as sandblasting the appropriate parts and applying one or more coats of primer prior to painting.

Unfortunately, vintage wheels and rims don't take kindly to age, especially if they were ballasted with

Surprisingly, the tires on this Model B, with their unusual tread, could be reused after new tubes were added.

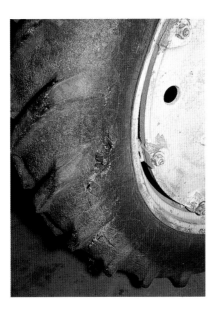

Although tire reliners can be used to salvage some tires, the hole in the sidewall on this rear tire has ended its usable life.

calcium chloride. They don't take well to rust, either. If one or more of the wheels are too far gone, or if a rim is in very bad shape, the best bet—particularly if you're dealing with a clamp-on rim—will be to frequent the auctions, swap meets, salvage yards, and classified ads for a replacement.

If you're only dealing with a few rust holes, however, you should be able to take care of those by thoroughly cleaning the holes and filling them in with small beads of weld or some type of filler, such as J-B Weld. You'll want to use a filler primer before painting to further smooth imperfections.

If you're good with a welder, you may be able to repair larger rust areas by totally replacing the damaged area. The first step will be cutting out the corroded or rusted portion of the wheel. Quite often, this will be the outer edge of the rim, where it has rested on the ground. Next, you'll have to find a scrap wheel or rim that is identical in size and style, from which you can cut a replacement piece. Make sure all edges have been ground smooth, clamp the splice into position, tack weld around the whole piece to keep it from warping, and carefully weld it into place. Once you've finished welding, grind all splices down until they are flush with the surrounding metal and prepare the wheel for painting.

Another option is to have new rims installed on your wheels. Dennis Funk, who restores John Deere two-cylinder tractors on his Hillsboro, Kansas, farm, says he has found that to be the best solution. Detwiler Tractor Parts in Spencer, Wisconsin, offers this service on several spoked and pressed steel wheels. Nielsen Spoke Wheel Repair in Estherville, Iowa, is another vendor that will put new rims on the centers of your old wheels. (See the appendix for addresses and phone numbers.)

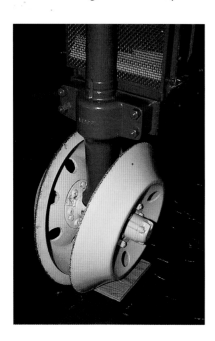  

John Deere used a variety of wheel sizes and types on its two-cylinder tractors over the years, which can lead to a lot of confusion about what wheels should be considered original.

Tires, Rims, and Wheels / 123

Skeleton wheels were just one of the steel wheel options John Deere offered on row-crop models. Many collectors find the thin wheels particularly appealing.

Another option on early steel-wheeled tractors was this wider wheel with bolt-on lugs.

The wheel extensions on this Model D's steel wheels were the next best thing to dual wheels.

When you find a tractor sitting on one or more rims, you'll need to carefully inspect the wheel for signs of rust or holes that might call for wheel replacement or a patch in the rim.

Above left: Believe it or not, steel wheels were still an option when this rare 1958 630 LP Standard was built. This tractor, which is on display at the John Deere Collectors Center, is one of only sixteen built in this configuration.

Above right: Having been cleaned and prepped, these steel wheels are ready for a coat of primer and paint.

Right: Once wheels have been sandblasted or cleaned, it's a good idea to hit them with a coat of primer to prevent any new rust until you're ready to reinstall them.

Both photos: For both safety and convenience, installing wheels on a tractor is often a two-person job.

CHAPTER 11

Hydraulic System

Hydraulic systems were a rare commodity on many early tractors. John Deere was the first to offer any kind of lift mechanism on a tractor when it introduced a mechanical power lift on the general-purpose Model GP in 1928. A few years later, when Deere introduced the Model A, the company was first again in offering a built-in hydraulic system on a wheeled tractor. The Model B subsequently became the second production tractor in the industry to have an integral hydraulic lift system.

Over the course of production, Deere used two different hydraulic systems, which were vastly different from one another. The Power Lift version, which is found on H Series tractors and early production A, B, and G Series tractors, is operated by a heel pedal and provides lift capacity only. The weight of the implement or manual force is required to lower the three-point hitch unit. The operator doesn't even have the option of stopping the unit part way up; the three-point is either all the way up or all the way down. Part way through production, a selective control modification was offered, which allowed the operator to stop the three-point hitch in any raised position.

However, nothing could improve on the inconvenience associated with the PTO-driven pump. Because the pump is driven directly off the PTO shaft, you either have to be moving or you have to stop, shift into neutral, and let the clutch out again in order to raise the three-point hitch or an implement connected to the hydraulic remotes. Either way, the PTO must be engaged. Of course, you can lower the unit or implement at any time, since gravity does the work, although it was a cushioned drop.

Hydraulic System / 127

The cylinder, which powers the rockshaft on A, B, G, and H tractors equipped with the Power Lift system, extends into the final drive housing between the bull gears.

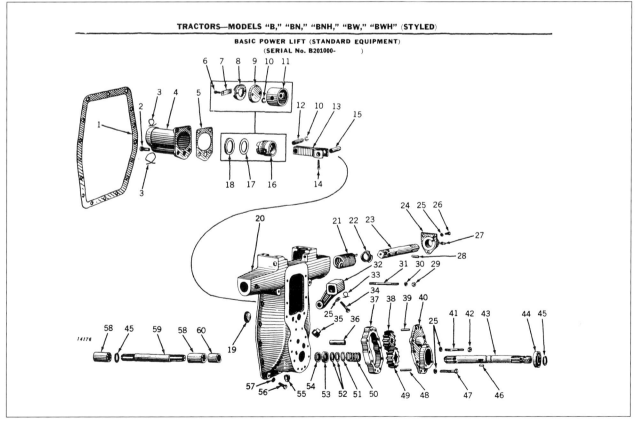

As illustrated in this exploded view drawing, the John Deere Power Lift system is not as complicated as it would seem. Note that the pump is driven directly off the power takeoff (PTO) shaft.

With the advent of the 50, 60, and 70 models, the hydraulic pump drive was moved to a housing on the back of the governor, where it was equipped with a control that could be used to disengage the pump on demand. This also allowed the use of "live" hydraulics.

The Powr-Trol system, meanwhile, is operated by a hand lever on the right side of the driver's seat. Introduced in 1947, it improved on the Power Lift system by providing both up and down pressure on the three-point hitch. It also allows the operator to stop movement in any position and provides two-way pressure at the hydraulic remotes. Best of all, the system used on most models is connected to a continuous drive, which means you don't have to have the clutch engaged to raise or lower an implement. The exceptions were the Model A, B, and G tractors equipped with Powr-Trol prior to the advent of a live PTO. However, there were again efforts by aftermarket vendors to correct any temporary shortcomings in Deere engineering. In this case, it was a kit that sandwiched a pump between the distributor and the governor to provide continuous pressure to the system, even if the PTO was not operating.

By the time the 50, 60, and 70 models came out, hydraulic pump drive was no longer the critical issue. For one thing, the hydraulic pump drive was moved to a housing on the back of the governor, where it was equipped with a control that could be used to disengage the pump on demand. However, with the advent of diesel models on the 20 Series, it was necessary to locate the pump elsewhere. Hence, 20 and 30 Series gas models continue to have the Powr-Trol pump attached to the governor housing, while diesel models have the pump located under the crankcase, where it is driven off the timing gear on the left side of the engine.

Most Dubuque-built tractors, meanwhile, have the hydraulic pump mounted on the engine timing gear cover, where it is driven via a coupling on the camshaft. The M and MT even used two different types of pumps—one with a three-piece body and an older version with a two-piece body.

Basic Principles

Before we look at troubleshooting and repairing the hydraulic system, let's look at some of the basics associated with hydraulic systems. First of all, hydraulic fluid is just like any other liquid—it has no shape of its own and acquires the shape of the container. Because of this, oil in the hydraulic system will flow in any direction and into any pump or cylinder it's directed, regardless of the size or shape.

Like any fluid, it is also practically incompressible. As a result, when force is applied to hydraulic fluid, it transfers force to the work site. Hydraulic fluid has one other characteristic: It has the ability to provide substantial increases in work force, meaning that one pound of pressure on the piston in a small pump is converted to several pounds on a larger cylinder or piston. If you use a hydraulic bottle jack to lift your tractor, you already know how this works.

Now, let's look at the application of this principle on your tractor. Instead of pressure being supplied by a jack handle and piston, it comes from a hydraulic pump. From there, it flows through a valve, which directs its path to the appropriate point, and finally to the hydraulic cylinder that rotates the rockshaft on the three-point hitch, power steering, or remote coupler.

Although hydraulic pumps used on farm tractors today are one of three types—gear pumps, vane pumps, and piston pumps—all of the hydraulic pumps used on John Deere two-cylinder tractors were of the gear type. That includes the power steering pump, which is separate from the Power Lift or Powr-Trol pump.

Although the pump on the fan shaft provides hydraulic pressure only to the power steering system, it should be inspected and repaired in the same manner as the main hydraulic pump.

Contamination

The two biggest enemies of a hydraulic system are dirt and water. If these two contaminants were kept out of the system by the previous owner, you may not have any problems with the hydraulic system. But if they managed to work their way into the system, you may have some repairs ahead of you.

Just as in the engine, dirt can score the insides of cylinders, spool valves, and pumps. Water, meanwhile, will break down the inhibitors in the hydraulic oil, causing it to emulsify and lose its lubricating ability—again, leading to scoring of cylinder walls and breakdown of internal seals. Unfortunately, the tolerances in many hydraulic pumps and spools are even tighter than those in engines.

One of the main ways dirt enters a system is through the air breather in the reservoir. The air breather is designed to let air move in and out of the reservoir in response to changes in the fluid level. It is also supposed to screen out dust by trapping it in between layers of oil-saturated filter material. Unfortunately, many farmers failed to clean the filter or check for cracks or leaks that permitted dirty air to penetrate the system. (Keep in mind, however, that many older tractors do not have a special reservoir or cooler for the hydraulic system, but simply use the same oil that lubricates the transmission and differential.)

Another way dirt can get into the system is through the careless handling of the hoses, particularly if a broken or damaged hose is replaced. This should be a hint to practice cleanliness when making hydraulic repairs or inspections.

Dirt isn't the only enemy, though. Sludge, which is formed by the chemical reaction of hydraulic fluid to excessive temperature changes or condensation, can also cause havoc. If enough sludge builds up on the pump's internal parts, it will eventually plug the pump. To add insult to injury, a restriction on the inlet side of the pump can starve it of fluid, and heat and friction will cause the pump parts to seize.

Troubleshooting

Like many other systems on the tractor, you should have an idea whether the hydraulic system actually works and how well it works from your initial test drive, assuming that the tractor was in running condition when you acquired it. Did the three-point hitch system raise and lower properly? Would the hydraulic remotes adequately raise an implement?

If your tests indicate a weakness in the system, the first thing you should do is check the fluid level in the reservoir and make sure it is filled to the proper level with the recommended grade and type of fluid. Improper fluid can not only cause low or erratic pressure, but it also can eventually deteriorate seals and packing, particularly if it contains incompatible ingredients.

Next, check for any problems with the hoses, including kinks or leaks. Keep in mind that hydraulic pressure escaping under high pressure through a pinhole leak can actually penetrate the skin, leading to gangrene poisoning if not treated quickly. Hence, you should never check for leaks with your bare hands, or even with leather gloves. Instead, pass a piece of cardboard or wood over any suspected areas to check for escaping fluid.

If you have access to a pressure gauge or know someone who can assist, you can also check the pressure in the system to see if the problem is in the pump or in one of the valves. On a Waterloo-built tractor, this can be as simple as connecting a gauge equipped with a shut-off valve to the remote outlet valves. On other tractors, such as the M or MT, it requires the removal of a plug in the valve housing. Check your service manual for the correct procedure.

If the problem is traced back to the pump, you can either locate a replacement pump or try to rebuild the pump yourself, realizing the need for absolute cleanliness.

Finally, refer to your service manual for procedures on how to check and adjust any by-pass valves, relief valves, and operating valves found in the system. Most valves utilize a spring that must meet a specified free length or compression rating to be effective.

The single remote cylinder valve housing on this Waterloo-built two-cylinder model is located where it is easy to check pressures and fittings.

This top view of the rockshaft housing on a typical Waterloo-built tractor, shows the rockshaft, rockshaft operating cylinder, and associated parts.

Hydraulic Seals

Due to the high pressure within the hydraulic system, no circuit can operate without the proper seals to hold the fluid under pressure. Seals also serve to keep dirt and water out of the system. Consequently, most problems associated with the hydraulic system—those that cannot be attributed to a defective valve or faulty pump—can be traced to a leaking or worn seal.

In general, hydraulic seals fall into one of two categories: static seals that seal fixed parts and dynamic seals that seal moving parts. Static seals include gaskets, O-rings, and packings used around valves, between fittings, and between pump sections. Dynamic seals include shaft and rod seals on hydraulic cylinder pistons and piston rods.

As for seal types, they can include O-rings, U- and V-packings, spring-loaded lip seals, cup and flange packings, mechanical seals, metallic seals, and compression packings and gaskets. Troubleshooting, naturally, consists of looking for leaks. Even though the perfect seal should prevent all leakage, this is not always practical or desirable. In dynamic uses, for instance, a slight amount of leakage is needed to provide lubrication to moving parts.

On the other hand, internal leakage, either from static seals or excessive leakage from dynamic seals, is hard to detect. Often, excessive leakage from an internal seal must be indicated by other means, such as pressure testing.

As a general rule, it's usually best to replace all seals that are disturbed during repair of the hydraulic system—assuming they are available. As is the case with the engine, transmission, and most other major components, it's a lot cheaper to replace a few seals or gaskets during restoration than come back later and do a repair job to correct leaks.

It may sound a little extreme, but you should give seals the same care during handling and replacement as precision bearings. This means keeping them protected in their containers until you're ready to use them and storing them in a cool, dry place free of dirt.

Following installation, static O-rings used as gaskets should be tightened a second time, after the unit has been warmed up and cycled a few times, to make sure they seal properly.

Dynamic O-rings, on the other hand, should be cycled or moved back and forth (as on a hydraulic cylinder) several times to allow the ring to rotate and assume a neutral position. In the process of rotating, the O-ring should allow a very small amount of fluid to pass. This is normal, since it permits a lubricating film of oil to pass between the O-ring and the shaft.

When rebuilding any type of hydraulic pump, it's a good idea to replace all bushings and seals.

Chapter 12

Electrical System

Unless you're restoring a late-model two-cylinder, there's not much to worry about on the electrical system. That's because there wasn't one—at least not as we think of it today. The closest thing to an electrical system was the magneto, which is essentially a generator, coil, and distributor wrapped up in one unit. Not until after World War II did Deere begin putting electric starters and lights on tractors as standard equipment. These, of course, called for a battery, a charging system, and ultimately, the switch to a distributor.

Magneto Systems

Depending upon the model of your two-cylinder John Deere tractor, it will likely be equipped with one of three different magnetos. Model D, H Series, and early production A, B, and G Series tractors were equipped with either an Edison-Splitdorf, Fairbanks-Morse, or Wico unit.

Basically, a magneto works much like a miniature spring-driven generator. As the drive cog is rotated, the magneto drive wraps up a spring that is positioned between the drive link and the magneto rotor. At the appropriate moment, the spring is released and the rotor is quickly rotated within a magnetic field to generate a charge of electricity—hence, the clicking sound associated with magneto operation. The distributor portion of the magneto determines which of the two cylinders receives the spark.

Due to the complexity of the spring-drive system and the internal workings of a magneto, it's best to have any adjustment or rebuilding done by a professional shop; there are several listed in the appendix of this book. However, there are some things you can check and replace yourself, including the points, condenser, rotor cap, and rotor tower cap.

You'll also need to make sure the magneto is reinstalled properly. This is especially important, since the magneto must initiate the spark before the piston reaches the top of the compression stroke.

On the other hand, the impulse coupling is designed to retard the timing during slow engine revolution, such as when it is being started. The result is a stronger spark for starting the engine and a reduced chance of kick-back, which can occur when the spark reaches the cylinder before the piston has reached the top of the cylinder. If you do not hear the click, your magneto may have a broken impulse coupler. Options for fixing this are to buy a new magneto or send the magneto to one of the businesses that specialize in restoration.

Since they required no electrical power, a magneto was used on almost all early model tractors to provide spark to the cylinders. Early models, like this D, with its brass-clad magneto, are particularly appealing to collectors.

Unlike modern electrical systems, the magneto operates like a generator, coil, and distributor wrapped up in one unit, which is generally operated off the governor shaft.

If the rotor magnet on a magneto gets too weak, most professional rebuilders have the equipment to restore the magnetic fields.

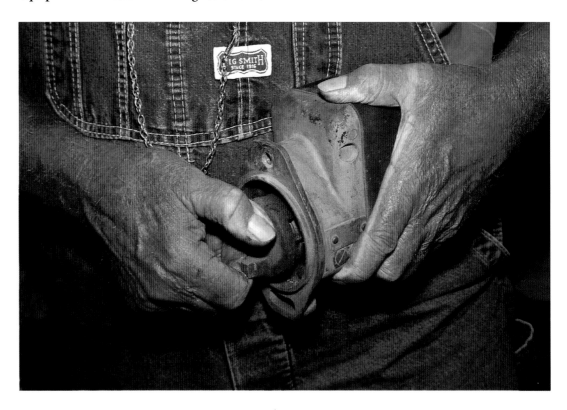

By carefully turning the pawl on the drive end of the magneto, you can listen for the click that indicates a spark is being created. By attaching a wire between the spark plug wire outlet and the case, you can also test for a spark.

When overhauling a magneto, it's always a good idea to replace the points and condenser.

Carefully check the cover and rotor tower for cracks or defects. Voltage leaks can ground out the unit and make it ineffective.

Magneto Inspection and Service

While there are a couple of ways to check for electrical output, one of the easiest for a novice is to attach a spark plug, via a piece of electrical wiring, to the coil output terminal. Then ground the spark plug to the base of the magneto and test for a spark while rotating the driven lug. A word of warning, though: Keep your hand clear of the coil and coil output terminal. If it's working properly the coil can put out nearly 30,000 volts, which you will obviously feel.

Most auto parts stores carry a spark tester that makes testing even easier. Basically, a spark tester looks like a spark plug with a large alligator clip. Hook the device to a plug wire and connect the clip to ground. When the tractor is cranked, the tester will flash if you have sufficient spark.

The presence of a spark won't be a guarantee that the magneto is putting out enough current, but it can give you an idea of how well the unit is working and if the spark is hot enough for ignition. If there isn't any spark, you at least know you're wasting your energy. The other testing alternative is connecting a special multimeter that records and stores readings in excess of 30,000 volts.

If you do choose to disassemble the magneto and try to service it on your own, the first thing you should do is inspect all the gaskets and insulators that isolate the generating components from the body of the unit. This especially applies to any bolts that protrude through the magneto body. Any voltage leakage can ground out the unit and make it ineffective.

You'll also want to clean all dirt and grime out of the unit, using a combination of compressed air and electrical parts cleaner. Finally, inspect and replace, if necessary, any seals or bearings that appear to be faulty. While you have the unit apart, it's also a good idea to have someone recharge the large horseshoe-shaped magnet that generates the magnetic field around the armature.

Before putting everything back together, apply a light coat of oil to all drive parts and bushings. Be careful not to over-apply the oil, though. Lastly, replace any tune-up components for which you can find parts. This includes the condenser, points, rotor tower, and rotor cap. It's easier and cheaper to replace them now than struggle with problems later on.

Timing the Magneto

There are two phases to magneto timing. The first step is to get the magneto roughly timed. The second is to get it adjusted for final timing.

To begin the timing process, the engine should be at the top of the compression stroke of the Number 1 piston. On horizontal two-cylinder John Deere engines, this is the left cylinder as viewed from the driver's seat. Start by removing the spark plug on the flywheel side of the tractor. Now, place your finger over the hole and rotate the flywheel in the direction it normally operates until you feel air pressure pushing out of the cylinder. If you need both hands to turn the flywheel, simply place a piece of tissue over the spark plug hole and watch for the air pressure to push the tissue aside.

When air starts coming out, it means you are headed into the compression stroke. At this point, continue to slowly turn the flywheel until the "L.H. Impulse" mark, which is stamped on the flywheel, lines up with the timing mark on the tractor frame. In this position, the slot in the governor shaft coupling should be horizontal.

Now, line up the driving lug on the magneto. To do this, hold the magneto in the same upright position as when it is mounted on the tractor. Next, insert the end of a short piece of wire into the upper terminal of the distributor cap. Bend the other end around to within $1/8$ inch of a metal part of the magento frame, cutting it off if necessary.

Finally, grasp the driving lug on the impulse coupler and turn the impulse in a counterclockwise direction (to the left). Stop turning the second the impulse trips and a spark shoots between the end of the wire and the magneto frame. The driving lugs on the impulse coupling should now be in a horizontal position that matches that of the governor shaft.

Carefully install the magneto and gasket on the governor case, making sure the impulse coupling lugs engage the slots on the governor drive coupling.

As soon as you have tightened the mounting bolts finger tight, rotate the body of the magneto as far as possible in the opposite direction of normal rotation. In most cases, this will be counterclockwise toward the front of the tractor. Now, go back to the other side and rotate the flywheel one complete turn and line the flywheel mark "L.H. Impulse" exactly with the timing mark on the tractor frame.

Finally, slowly tap the top of the magneto toward the rear of the tractor until the impulse trips. This indicates that the magneto is right at the point where it will fire the Number 1 piston. Tighten up the bolts and install the spark plugs and plug wires, making sure the left-hand spark plug cable is installed in the upper distributor cap terminal.

Final timing involves carefully rotating the magneto by hand until it has been correctly positioned, then tightening the mounting bolts.

A set of marks on the flywheel serve as a guide when reinstalling the magneto or distributor and timing the engine on horizontal engine models.

Distributor Inspection and Repair

In the majority of cases, John Deere two-cylinder tractors equipped with a distributor utilized one of two brands. Although most later models used a Delco-Remy unit, earlier tractors—such as the styled A, B, and G—used a Wico distributor. Both are relatively simple, in comparison to a magneto, since the distributor doesn't actually produce electricity, but only serves to distribute the spark to the appropriate cylinder. Hence, service and rebuilding are much easier, as well.

Unless you plan to replace the spark plug wires, start distributor restoration by grasping the spark plug and coil wires and gently twisting the boot as you remove them from the cap. It's also a good idea to label the wires with tape and a marker, so they can be reinstalled in the correct position.

If you plan to do much work on the distributor or engine, you may want to go ahead and remove the whole distributor at this time and clamp it in a vise where it will be easier to work on. Getting it reinstalled and timed correctly isn't that much different than reinstalling a magneto. Before you throw the cap in the trash, check for carbon tracking around the plug towers and around the cap base. Defective spark plugs or spark plug wires can cause sparks to travel from the tower to the nearest ground, which is usually the mounting clip. The evidence is a small carbon trail that resembles a tiny tree root.

Once you have inspected and cleaned all parts at the top end of the distributor, it's time to take a look at the bottom end. Start by checking for any free play or wobble in the drive shaft. Generally, there are two bushings or seals on the shaft—one at the top and one at the bottom. If either one is badly worn, it should be replaced as part of the rebuild.

Next, remove the points and condenser. Beneath the mounting plate you'll find a spark advance system of springs and counterweights. Replace any springs that are broken or weak; make sure the weights aren't rusted and that they move freely. The springs must be replaced in pairs and by springs of identical size. Clean all pieces in solvent before reassembling.

Although some restorers like to reuse the points, touching them up with a file before they're reinstalled, a quality rebuild should include new points, condenser, rotor, cap, and spark plugs.

Setting the Point Gap and Timing the Distributor

You will need to adjust the gap between the points. To do this, rotate the shaft to a point where the cam lobe separates the points, creating a gap. Now, using a wire-type feeler gauge, adjust the points to match the specifications in your service manual and tighten the mounting screws.

Be sure you remove any oil film from the feeler gauge before inserting it between the points. Oil on the contact points can cause the points to burn or become pitted. Finish off point adjustment by lubricating the rubbing block with a small amount of high-temperature grease.

At this point, you can install a new rotor, line up the casing marks, and reinstall the distributor in the housing. If you've compensated for the gear taper during installation, the mark on the distributor casing should line up with the rotor tang. If not, it shouldn't be off by more than one tooth.

If the distributor is completely off, though, there's no need to panic. Most service manuals include a detailed procedure for timing the engine. You can also retime the engine by placing your finger or a tissue over the Number 1 spark plug hole, just as you would when timing a magneto. Then crank the engine until the Number 1 cylinder begins compression. Continue turning the engine until the timing mark on the flywheel lines up the with the timing mark or indicator pin.

Now, turn the drive shaft on the distributor until the rotor lines up with the terminal for the Number 1 spark plug. Hopefully, you marked the distributor casing earlier, making this an easy task. Offset the rotor a little to compensate for the gear mesh and slide it in. Some engines will have a tang at the bottom of the distributor drive shaft that runs the oil pump. Make sure this tang engages the pump shaft. If it doesn't, put a little downward pressure on the shaft and rotate the engine a quarter or half turn. It should drop into position. Don't tighten the bolts yet, though.

Electrical System / 139

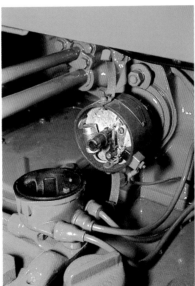

Rebuilding a distributor is not nearly as difficult as overhauling a magneto, since a distributor doesn't generate its own electricity.

Check the inside and outside of the distributor cap and individual cap towers for cracks, burned spots, and corrosion. If there is any doubt about the condition, replace the cap.

A quality distributor rebuild should include new points, condenser, rotor, and spark plugs.

As was the case with the magneto, you'll need to make sure final adjustments have been made before tightening the brackets that hold the distributor in place.

Install a new distributor cap and spark plug wires and finish the timing process with the ignition on. To do so, slowly turn the distributor in the direction of normal rotation and watch for the exact moment that a spark occurs at the plug. If you missed it, back the distributor up and try it again. The engine should now be timed properly. Tighten the distributor mounting bolt(s). You may still want to check everything with a timing light, if you have one available.

Coils

Perhaps the easiest way to check the coil is to gently pull the wire that runs from the coil to the distributor and hold it about 1/8 inch from the engine block or a good ground. A strong spark should jump the gap when the engine is turned over. The coil should also be clean and dry. If you're not confident that it is in good condition, consider replacing it.

Generators and Voltage Regulators

If you ever did any experiments with electricity or with a generator in high school science or physics class, you may already have an idea how the generator on your tractor works. But if not, don't despair. It's not that complicated.

By the simplest explanation, one way electricity can be created is by moving a conductor through a magnetic field. So if you look at this principle in terms of a generator, the armature, which serves as the conductor, is moved, or in this case spun, inside of two or more magnets. But think back again to science class. Remember the time you wrapped electrical wire around a nail and attached it to a battery? You created an electromagnet. The generator on your tractor just uses a larger version.

So now you can envision the field coils as electromagnets attached to the generator case. As the armature spins within this magnetic field, electricity moves through the armature to where it is allowed to flow though the brushes.

The wire that makes up the coils actually begins at the F terminal of the generator, winds its way around the case, and terminates either at a third brush on a three-brush generator or connects to the wire going to the output brush or A terminal on a two-brush generator. Unless the generator has been replaced, it most likely uses two coils.

The magnets are actually a two-piece arrangement consisting of a coil of wire that fits around a pole shoe made of a special kind of metal. As the armature spins within the magnetic field, it begins to generate electricity—some of which is used to charge the battery or run electric lights, while the rest goes back into the field coils to make the magnetic field even stronger. The role of the regulator is to control the generator by manipulating the ground connection of the field coils.

Among the most important components are the brushes, which serve to gather the electricity that is being produced. The brushes ride on the commutator and allow the generated electricity to travel to the voltage regulator, and ultimately the battery or other load, then back into the field coils.

As a result, the primary problem areas on a generator are the brushes, commutator, and bushings. Generally, brushes should be replaced if they are worn more than halfway. Quite often the bushings and bearings that support the main shaft will also be worn and require replacement.

To perform any of these repairs, however, it will be necessary to disassemble, clean, and inspect the generator. Don't try to remove the field coils unless it is absolutely necessary. If you take them out, it will be difficult to get them back in without the proper tools. There are only two things that can go wrong with the field coils anyway—either the wires have lost their insulation somewhere and are touching ground (such as the generator casing), or they have broken and have created an open circuit.

Considering the cost and availability of rebuilt generators, most restorers will tell you it's seldom worth your time to try to repair a generator yourself. Moreover, most communities have a machine shop or automotive shop that can test and rebuild your generator to factory specifications.

The same could be said about the voltage regulator. If there is doubt about whether it is working properly, your best bet is to take it to a shop that specializes in starter and generator rebuilds and let them test it. If it can be repaired, they'll be able to take care of it, and if not, they should be able to suggest the correct replacement.

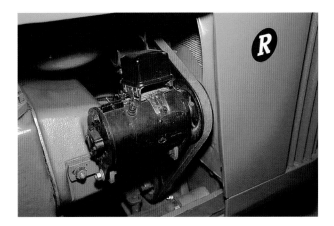

Both photos: **The function of the generator is to replace any electricity in the battery that has been used elsewhere in the electrical circuit. Hence, generators weren't needed until the advent of lights and electric starters on tractors.**

Any generator rebuild usually includes putting the armature on a lathe and turning the commutator, or at least polishing it with a piece of emery cloth.

The brushes on generators and starters alike should be replaced if they are worn more than halfway.

If you're overhauling a starter or generator, it's worth the time and money to put in new bearings.

Starters

Not every two-cylinder tractor is going to have a starter. Even when electric starters became available, they were only an option on many models.

In essence, the starter is much like the generator, only it operates in the opposite manner. Instead of generating electricity, it takes electricity and uses it to turn a drive sprocket. However, like the generator, it contains an armature, coils, brushes, and commutator that can wear or short out in much the same manner. Hence, overhaul consists of inspecting the brushes for good contact with the commutator and making sure the latter is reasonably clean and smooth. If it is not, it will need to be turned down on a lathe.

Just as you did with the generator, you'll also need to check for worn, dirty, or damaged bearings. Again, it may be easier and less costly in the long run to have these things done by a shop that specializes in starter and generator repairs or trade it for a rebuilt unit.

Before you pull the starter off the tractor, though, you need to realize that some tractors have a neutral start interlock switch that prevents the tractor from being started if the tractor is in gear. If the interlock switch is faulty, you may be wasting your time on the starter itself.

You may also have problems related to the battery or loose or corroded connections. To check for these, connect a fully charged battery to the starter using a set of jumper cables. If there is a significant difference in the way the starter turns over, check the battery; then inspect, clean, and tighten all starter relay connections as well as the battery ground on the frame and engine. If the starter still does not crank with the jumper cables, plan on removing and replacing or repairing the starter.

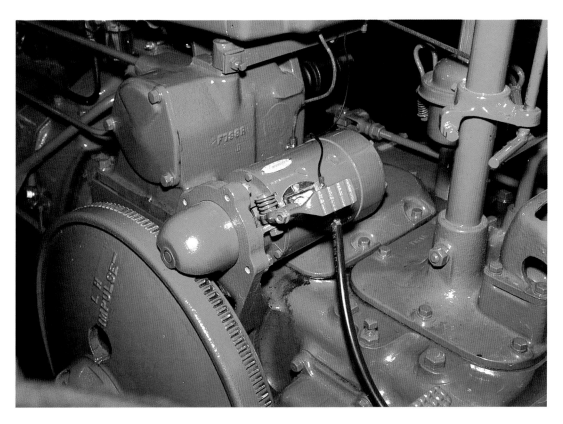

The starter on early two-cylinder models was pretty basic. With no cover and only a simple foot switch for engagement, it made direct contact with the flywheel.

Wiring

Unless your tractor has been treated with tender loving care for the past fifty to seventy-five years, it's doubtful that you will get by without replacing at least part of the wiring. At the very least, you will want to replace the spark plug wires as part of the engine rebuild.

However, replacing crimped, spliced, and inferior wiring on the rest of the tractor not only improves the looks of your restored tractor, but it also can be a safety measure. Wiring with cracked insulation and wires on which the old, cotton braiding has dry rotted can ground out electrical components—and at the worst, create a fire hazard.

The first step in wiring restoration is locating a wiring diagram. Hopefully, this will be included in your service and repair manual. If not, you'll have to trace the wiring from the power source, or the battery, to each switch and component. Don't put too much stock in what you find, though. Previous owners may have replaced the original wiring with a different gauge of wire, the incorrect type of wire, or they may have even taken shortcuts with the routing.

If there is any doubt, talk to other tractor owners or try to find a well-restored model like your own and take notes. As a general rule, you should use at least 10-gauge wiring for circuits that carry a heavy load, such as from the generator. Switches and other components can be wired with 14-gauge wire. Remember, the larger the gauge number, the smaller the wire diameter.

If you have a lot of wiring to replace, it might be easier to just purchase a complete wiring harness for your tractor. Available through various sources, such as those listed in the appendix, an appropriate wiring harness is made up with the correct gauge and color of wiring for each switch, gauge, and component and is pre-wrapped to match the routing.

Finally, you'll need to consider what role historical accuracy plays in your restoration goals. If you're restoring the tractor as a working machine, you can get by with modern automotive wiring in the correct gauge and crimp-style connections. However, if you're going for an accurate restoration, you'll need to locate the appropriate gauge of lacquer-coated, cotton-braided wiring for any tractor that used cotton-covered wire as original equipment. Generally, cotton-braided wiring is appropriate for any tractor built before World War II.

The use of lacquer-coated, cotton-braided wiring—which was common on tractors built before the mid 1950s—really sets off an early model restoration.

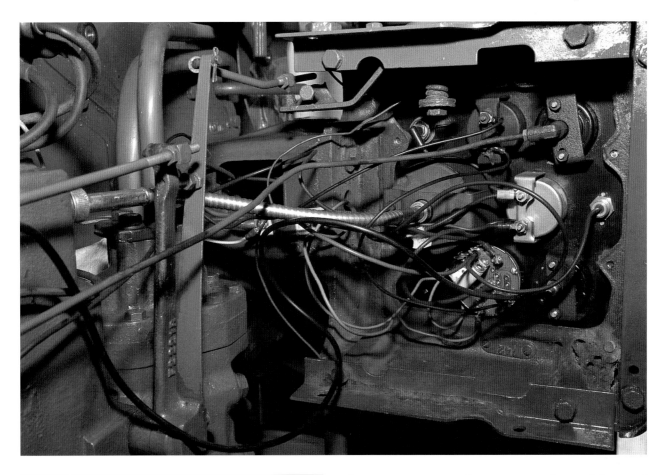

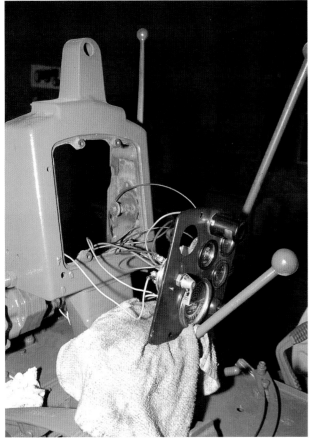

Some restorers prefer to use modern wiring and connectors to restore the electrical system, particularly in areas where it isn't visible.

Spark Plug Wires

Contrary to their rugged appearance, spark plug wires are quite sensitive. That's because most secondary wires consist of a soft-copper-core wire surrounded by stainless steel or carbon-impregnated thread mixed with an elastomer-type conductor. The outside covering of a heavy layer of insulation prevents the 12,000 to 25,000 volts from bleeding out when the wire is carrying current.

Consequently, the resistance-type wires do not handle sharp bending or jerking and can break internally, ruining the wire. Excessive exposure to oil and antifreeze can chemically break down the coating, as well. Perhaps the most damaging effect is caused when someone pulls on the wires to remove them from the spark plugs. This can separate the conducting material, causing internal arcing.

Therefore, it's important that you keep spark plug wires clean and separated from each other with any wire clips that are integral to the routing. To remove wires from the plugs during service or restoration, grasp the rubber boot, not the wires. Keep in mind, too, that secondary wires, including the coil wire and spark plug wires, can appear in good condition, yet be faulty.

One way to check their condition is to measure the wire resistance using an ohmmeter. In general, the wires should register around 8,000 to 12,000 ohms per foot. This will also test the wire for continuity, ensuring that there are no breaks in the copper-wire core.

Lights

Depending upon the age of your tractor, you may not even have to concern yourself with refurbishing the lights. Lighting systems were available from John Deere in the 1920s on the D and GP tractors. Model A and B tractors were offered with electric lighting as an option from the start of production. However, lights on tractors didn't become standard equipment until the late 1930s.

Besides broken lenses and deteriorated wiring, the most common problem you're likely to encounter is a rusty, faded, or worn reflector. Most restorers start the renewal process by disassembling the light, then cleaning it inside and out. One restorer likes to bead blast the light housing to a smooth finish—although he admits paint stripper can have the same effect.

Then, it's simply a matter of using a galvanizing-effect paint to spray paint the reflector. The outer shell can be painted at the same time you're painting other tractor components. Should you have any trouble finding a replacement gasket for reassembly, you might also want to use a tip provided by Jeff Gravert, a tractor restorer from Central City, Nebraska. He says a good substitute for the gasket that fits between the lens and the reflector is a strip of caulking that comes in rolls and pulls off like a piece of cord. The caulk also helps hold the lens in place.

Should you need to find a replacement light, swap meets, salvage yards, and dealer parts counters are all good sources. There are also a number of vendors that offer both reproduction and refurbished lights for sale.

The owner of this old John Deere two-cylinder tractor was lucky to find a model on which both lights were complete and unbroken.

Having been cleaned, painted, and fitted with a new rubber seal, this light is ready for reassembly.

Gauges

Ten years ago, gauges were a real problem for the John Deere restorer trying to obtain an authentic look. Reproduction gauges like the ones originally used on most John Deere two-cylinder tractors were unavailable, due to copyright restrictions. Unlike the replacement gauges currently available from John Deere dealers, which feature black faces, the original two-cylinder gauges had white faces and the words "John Deere" on the face. Since John Deere loosened the copyright restrictions, though, the correct gauges are now readily available from several sources.

On the other hand, if you're not concerned about authenticity, or just want to use your restoration as a work tractor, the black-faced gauges available through your local John Deere dealer will work fine.

Another option, of course, is to look for a replacement gauge at a salvage yard. Depending on the reason the tractor ended up there, it may have a gauge that's in better shape than the one on your tractor. If the face plate is in relatively good shape, it's easy enough to refurbish the rest of the gauge.

First, you'll need to carefully remove the bezel ring that holds the glass in place. On some gauges, you may have to bend up the lip around the edge of the gauge to do so. Now, it's just a matter of cleaning it up, making sure the mechanisms work properly, and repainting it.

The black-faced gauges on this John Deere two-cylinder tractor are not original. For an accurate restoration, they should be white; but white gauges are available only from reproduction parts suppliers. If authenticity is not important, you can get the black gauges from any John Deere dealer.

If the white gauges on early model tractors aren't reusable, you can get aftermarket replacements from a number of parts sources.

Chapter 13

Fuel System

There can be a wide variation in the amount of work a fuel system is going to need, depending upon the age of the tractor and whether it was running at the time you bought it. If you had a chance to drive it before the purchase or before you started tearing it down, you should have a good idea about how smoothly it was running.

On the other hand, if you're restoring a treasure that has been sitting in the weeds or an old barn for the last twenty years, chances are there's a lot of rust, varnish, water, and who knows what else in the system.

John Deere was rather unique in that the company offered customers a choice of fuel type all the way through the end of two-cylinder tractor production. Naturally, the first models were all-fuel only. However, by the time the numbered series made its debut, customers could choose from gasoline, all-fuel, liquefied petroleum (LP), or diesel models.

Fuel Tank

You can rebuild the carburetor, clean or replace the fuel lines, and change the fuel filter, but all that does little good if the fuel tank was the source of contamination. So the first step in fuel system overhaul should be to clean the tank and reseal the interior if necessary.

Everyone seems to have their own story about how to clean a fuel tank. Some have been known to stick the sandblast nozzle in the fill opening and move it around to hit all sides with silica sand or glass beads. The risk, of course, is that you might just blow a hole through any weak spot in the tank. Plus, you'll need to get all the sand out of the tank.

A more common option is to fill the tank about one-third full of water and add a few handfuls of ½-inch nuts, shingle nails, or pebbles. The key is to provide a slight abrasive action to clean up the inside. A word of caution is in order, though, particularly if you want to avoid frustrations later on. Check to see if the filler neck extends into the tank to act as a baffle that keeps fuel from splashing back out. If it does and you add a non-metallic abrasive like pebbles or even brass nuts, you may have a hard time getting all the material out of the tank. It will be a little like trying to shake pennies out of the slot in a piggy bank. Most professional restorers prefer nuts and bolts, because the stragglers can be fished out with a magnet.

The next step is to agitate the tank rather vigorously with this mixture sealed inside. One restorer claims the best way to agitate the mixture, assuming you do some farming, is to strap the tank to a tractor wheel with bungee cords and let it rotate while you do a day's field work. Another restorer says he does the same thing, but simply blocks the front wheels on a tractor, locks the brakes, and jacks up one rear wheel to which the tank is strapped. Then, he lets the tractor idle in gear for four or five hours, letting the tractor wheel work much like a rock tumbler. Yet another restorer says he secures the tank in the back of his pickup and hauls it around for about a month while he's working on other parts of the tractor. If you're going to use that method, it helps if you live on country roads. Anyway, you get the picture. You need to agitate the tank vigorously enough and long enough to scour all the rust and residue out of the tank.

If the tank is in really bad shape, you might want to start by adding a lye-based cleaner to the initial mixture for the first fifteen or twenty minutes and then switching to a clear-water-and-abrasive mixture. Once the tank has agitated for a sufficient period of time, remove the abrasive material (pebbles, nails, or nuts) and rinse the tank with clean water. You may have to repeat the rinsing process several times until you get clean water coming out of the tank.

If you find that the fuel tank leaks, do not try to solder it yourself. Regardless of what kind of instructions your friends have given you—like filling the tank with exhaust gas from your car's exhaust pipe, which supposedly makes it safer—it is impossible to get the fuel tank clean enough to solder safely in a home shop.

A vital step in fuel system overhaul is cleaning the tank and, if necessary, resealing the interior with a good quality tank sealer.

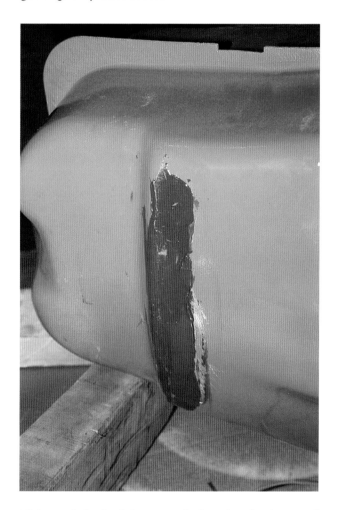

This tank leaked from pinholes that had rusted through the tank above the mounting brackets. Luckily, the holes were small enough to patch with a quality sealer.

Shops that repair automobile gasoline tanks usually steam clean the insides for an hour or more to ensure that no residual gasoline is emitted from the pores in the metal during heating. Even then, soldering a gas tank can be a dangerous proposition. That's why many professionals also fill the tank with an inert gas or liquid before heating the tank.

One thing you can try on your own is patching the hole with an epoxy, assuming the patch will be hidden beneath the tractor sheet metal. Several restorers have reported success with materials marketed as Magic Metal, J-B Weld, and other "gas tank menders" sold in automotive stores. The key is getting the surface clean with a good parts cleaner prior to mixing and applying the epoxy.

Once the tank has been repaired or proven to be free of leaks, it still makes good sense to coat the interior with a fuel tank sealer. One restorer, who shall remain nameless, got a tractor completely back together and painted, only to put five gallons of gasoline in the tank and discover a leak in the fuel system. He eventually traced it to a few pinholes where the fuel tank rested on the support straps that hold it to the underside of the hood. This is a spot that deserves extra close inspection on many John Deere models.

Most sealer formulas recommend that you first etch the tank with phosphoric acid or an acid metal-prep solution to stabilize any remaining rust prior to adding the sealer. Be sure to leave the fuel tank lid off, though, when rinsing the tank with acid, since the reaction with the metal creates a gas. Then rinse the tank several times with clean water and air dry the tank with a warm air source to prevent any further rust.

As soon as the tank interior has adequately dried, pour in enough sealer to cover all sides of the tank interior. In most cases, the sealer instructions will tell you to allow several days for the material to cure before adding fuel. While you're waiting, you can finish up the fuel delivery system by cleaning the sediment bowl assembly and replacing all gaskets and screens.

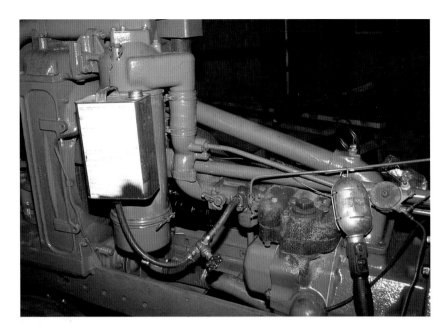

If you plan any engine work that involves starting and running the engine while the fuel tank is removed, you'll need to rig up some sort of temporary fuel supply.

Be sure to clean all fuel passages, including the inlet and outlet from the sediment bowl.

Fuel Hose Inspection

The fuel lines and fuel filter may look fine right now, but taking a few extra minutes to thoroughly examine them could still save you a lot of headaches later. Start by inspecting rubber fuel hoses for kinking, pinching at tight bends, and internal swelling. Also, make sure fuel lines are not running near an exhaust manifold or pipe, which could lead to a vaporizing problem on hot days. If the fuel turns into a gas in the line, it can cause the fuel circuit to "vapor lock" and stop delivering fuel to the carburetor.

Finally, make sure fuel is flowing freely to and from the fuel filter. A partially plugged fuel filter can lead to a leaner fuel mixture and cause backfiring, spitting, and misfiring. Unfortunately, when the engine dies, back pressure from expanding vapor can push debris from the filter back into the fuel tank, thereby hiding the problem. The engine will start and run like normal until debris, once again, finds its way back into the filter element.

It's important to note that John Deere began installing a hydraulically operated automatic fuel shut-off system shortly after the introduction of the first numbered series tractors. The shut-off is located on top of the fuel strainer on gasoline models and is operated by oil pressure that opens the fuel valve as soon as the starter is engaged and the engine turns over. If this valve is not working properly, you may drive yourself to exasperation trying to find the problem in the fuel lines or carburetor. Instead, it's possible that the rubber washer has deteriorated or the diaphragm has a hole in it, which allows fuel to run into the oil line.

Carburetor Repair

To your benefit as a tractor restorer, the carburetor on most vintage farm tractors is not as complex as it would appear. To begin with, there was no such thing as a fuel pump on most early tractors. The fuel tank was simply mounted above the engine and the fuel was fed to the carburetor by gravity. The carburetor itself is equally simple. John Deere used a Marvel-Schebler carburetor on nearly every model the company built. The exception was the use of a few Zenith carburetors on some early tractors. While most of the Waterloo-built tractors used a model DLTX unit, the Dubuque-built tractors used a model TSX carburetor.

Hence, adjustments in most cases are limited to the idle-mixture adjusting needle and load adjustment screw. Clockwise rotation of both the idle-adjustment needle and load-adjustment needle leans the mixture. Both needles are located on the top of the carburetor body. The idle-adjusting needle is brass plated, while the load needle is cadmium plated. On the dual-barrel carburetors, which were first used on numbered series Waterloo tractors, there is only one common load adjustment, but separate idle adjustments for each cylinder.

Internally, about the only adjustment that is ever needed is to bend the float stem to change the fuel level in the bowl. The float setting is important because the fuel level in the bowl plays a critical role in low- and high-speed adjustment.

In principle, the float bowl acts as a reservoir to hold a supply of fuel for the carburetor. However, it's important that the fuel in the bowl remain at a consistent depth, since the fuel level regulates fuel flow to the carburetor itself. As fuel fills the bowl by gravity, the float raises on a hinge and pushes the needle valve into a seat to shut off the fuel flow. In effect, it works in much the same way as the float and valve in the toilet tank in your bathroom.

If the fuel level in the bowl is too low, the engine does not respond readily when accelerated and it will be difficult to maintain carburetor adjustments. If the fuel level in the bowl is too high, it can cause excessive fuel consumption and crankcase dilution. Plus, it can cause the carburetor to leak. Again, it will be difficult to maintain carburetor adjustments.

Depending upon the carburetor model used on your tractor, the top of the float should either be ⅜ or ½ inch below the top of the bowl. In general, those models with a cast-iron fuel bowl have the ½-inch specification.

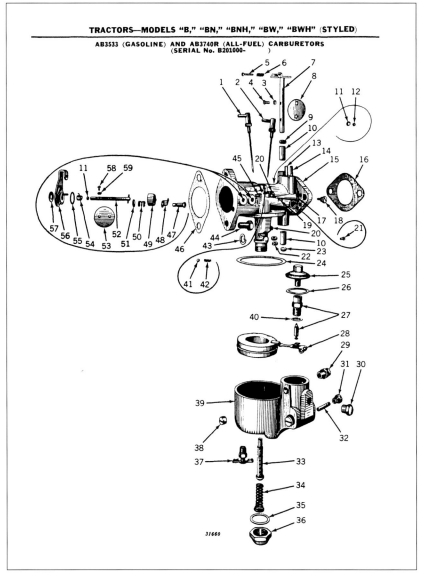

Above: Most carburetors are not as complicated as they might appear. If you don't think you can handle installation of a carburetor kit, though, there are plenty of businesses that offer rebuilding services.

Right: With the correct parts kit, a set of instructions, and an exploded-view drawing such as this, even a novice restorer can perform minor carburetor repairs.

Carburetor Rebuilding

The first step in rebuilding a carburetor is to remove it from the tractor and get it cleaned up. Start by closing the valve on the fuel tank, if this hasn't already been done or if the fuel tank hasn't already been removed, and disconnect the fuel line from the carburetor. You'll also need to disconnect any choke cables and governor linkage on most carburetors.

Now, remove the carburetor from the intake manifold and the air cleaner and move it to a clean workbench or other area where you can disassemble it without losing pieces. Remember that until you empty the bowl, the carburetor still contains a small amount of gasoline. So treat it as flammable until you've cleaned it out. If the carburetor is equipped with a drain plug or a drain valve, and if it's not rusted in place, it's best to drain the fuel before you go any further.

Begin by removing the top half of the carburetor and dumping any gasoline that is still in the bowl into a safe place. Next, carefully disassemble the carburetor, inspecting all parts for wear as you go. Before you remove the idle-mixture and load-adjusting needles, though, carefully tighten each against the bottom of the seat, noting how many turns it takes to do so. When reassembling the carburetor, you can again tighten the needles to the seat and back off the recorded number of turns. That will at least give you a starting point to dialing in the carburetor. If you have a good service manual, it will also tell you how many turns open are needed for initial settings. Of course, final adjustment must be made when the engine has been warmed up and is running.

Be sure you make notes—including notes about the orientation of any gaskets you remove—if you have any doubts about how the carburetor goes back together. Many of the parts will be replaced while installing a carburetor rebuild kit, but don't throw anything away until you know you have the proper replacement part. Some kits are applicable to more than one carburetor, so there may be parts you don't need—and that means you must be able to match the parts you *do* need!

Next, soak the two halves of the carburetor, along with the components you've removed, in a new container of carburetor cleaner for at least twelve hours or for the amount of time recommended on the label. Many carburetor cleaner solutions come with a parts bucket, so use it to turn and move the components from time to time.

Some restorers have also used a sandblasting cabinet and glass beads to scour the two halves of the unit once all the parts have been removed. You need to use care, though—and never use sand. Otherwise, you can quickly ruin the brass jets that remain in place.

You'll also need to clean these brass jets, either by removing them with a screwdriver or cleaning the passageway with a sturdy piece of nylon fishing line and an air hose, directing the air in the opposite direction as the fuel flow. Some jets are pressed into place and can't be removed without destroying them, which means you need to know ahead of time whether a replacement is available. If a jet is removable, make sure you have a screwdriver that fits the slot securely. Brass parts are easy to strip or damage.

Farm equipment dealerships and automotive parts stores are a good source of tractor parts, like this carburetor kit. Notice that it includes full instructions for carburetor overhaul.

The fuel jet must be clean and unobstructed for the carburetor to operate properly.

Soaking the disassembled carburetor in a bucket of cleaner should be the first step in the overhaul process.

If the tiny holes in this jet become plugged, it can dramatically affect the air-to-fuel mixture.

Carburetor Reassembly

Once the carburetor has been thoroughly cleaned, it's time to put it all back together and make the necessary adjustments. Hopefully, you were able to find a kit for your carburetor that contained all the appropriate parts.

It's important to inspect all moving parts and replace those that are damaged. If the throttle-shaft bushings or seals, for example, are worn to the point they are letting air leak into the carburetor, they're going to affect the gas-air mixture. In some cases, the throttle shaft itself may have a groove worn into it.

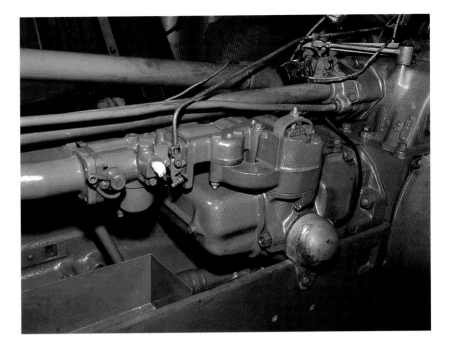

Be sure to replace any damaged hoses that connect the carburetor to the air intake. Air leaks can negatively affect the air-to-fuel mixture.

In the process of reassembling the carburetor, you also need to make sure all gaskets are properly oriented. Otherwise, you may block a vital orifice or passageway.

Also, when checking the float height, be sure the gasket has been positioned on the top half of the carburetor. The measurement specified in your service manual or kit instructions is almost always taken from the surface of the gasket to the top surface of the float.

Finally, when reinstalling the main-jet and idle-speed mixture screws, be sure to screw them all the way in and then back them out the number of turns recorded during disassembly, or as instructed in your service manual.

The groove worn into this throttle shaft allowed air to leak in and fuel to leak out, which negatively affected the carburetor's performance.

A new shaft should easily solve the problem.

Make sure all gaskets are oriented correctly when reassembling the carburetor. Otherwise, you may block a vital orifice or passageway.

When reassembling the carburetor, it's best to reset all adjusting screws and needles to their original settings as a starting point.

Diesel Systems

John Deere joined the diesel engine revolution in 1949, when the company introduced the Model R. Available as a standard-tread model only, the R required a 24.6-cubic-inch gasoline starter motor to get the main engine started.

The diesel engines used in modern farm tractors are relatively safe—not to mention immensely popular. However, they operate in an entirely different manner than gasoline engines. Because diesel fuel—also referred to at times as Number 2 fuel oil and distillate—is heavier and doesn't vaporize nearly as easily as gasoline, there is neither a carburetor nor spark plugs on a diesel engine.

Instead, diesel fuel is injected directly into the cylinder at pressures up to 2,500 pounds per square inch. In fact, some of today's diesel engines utilize pumps producing up to 5,000 psi pressure at the injector. It is this high injection pressure, combined with cylinder compression, that creates the heat needed to ignite the fuel. This means that the injector pump needs to force the precise amount of fuel into each cylinder, via an injector, at exactly the right time. That's why the injection pump is generally geared to the crankshaft.

That's also the reason injection-pump testing and rebuilding is best left to a professional who has the knowledge and the equipment to work on it. Roy Ritter, a Missouri-based tractor restorer who has been working on John Deere two-cylinder diesel pumps for nearly fifty years, says one thing you can do, though, is make sure the injector is the correct size for the tractor model and engine. As an example, he notes that every John Deere injector has a number stamped on the end. It may take a magnifying glass to read it, but it's there. Moreover, every John Deere diesel engine has a certain tip number specified.

"You wouldn't believe how often I receive a pump and injector set that has the wrong injector tips installed," he says. "Or one injector will have one tip number and the second one will have a another number. Once I've installed the right tips for their particular tractor model and cleaned the injectors, it's usually just a matter of getting the pump adjusted to deliver the correct pressure and volume."

Above: The injector pumps used on diesel engines sit above the engine, where they are driven by the camshaft.

Left: Because of the complexity of a diesel injector, it's important that the injector tip and shims match the tractor model and engine.

Oil-Bath Air Filters

For many of us, an air filter is a square or round element composed of aluminum screen and folded, paperlike material that traps dirt particles as air flows to the carburetor. Once it gets dirty or has been in place a certain length of time, it's tossed and replaced with a new one.

While some vintage tractors may have that type of filter, it's more likely that the engine on your tractor has an oil-bath air filter. After all, replaceable air filters didn't become a feature on farm tractors until the 1960s, when two-cylinder tractors were dropped from the John Deere line.

Let's first look at how the oil-bath air filter works. As you probably noticed, the filter itself looks like it is made out of a pile of metal shavings. What's more, there's a cup at the bottom filled with oil. When you start the engine, a certain amount of oil is sucked out of the oil cup or pan and onto the metal shavings, which then filter the air as it flows through the canister.

The filter canister is designed to be just the right height and size so that the engine can pull oil up into the screen along with the air, but without pulling it on into the carburetor and engine. If you notice, too, incoming air is drawn into the filter through a center pipe that leads to the bottom of the canister. As a result, any heavy dirt particles should fall directly into the oil cup. Lighter particles, of course, should be trapped on the oil-soaked filter surface as the air moves upward through the outer portion of the canister toward the carburetor.

Now that you understand how the filter works, it should also be easier to visualize the potential problems. The first comes with using the wrong weight oil. If you add oil that is too light, it can be drawn beyond the filter and into the engine. Using oil that is too heavy will have the opposite effect—not enough oil will be drawn up into the filter element, and much of the air-cleaning surface will go unused.

As inconvenient as it may sound, oil-bath air filters were designed to be cleaned and refilled daily when in use. In really dusty conditions, a farmer sometimes had to service the air cleaner a couple times a day. Naturally, the air cleaner is going to work best when the oil level is at the recommended level. However, letting the oil cup fill up with sludge can be even more detrimental. Simply adding more oil to the cup, in fact, can make it worse. When the particles-to-oil ratio gets to a certain level, the dirt will begin to hang onto the cleaning surfaces. Eventually, instead of just clean air being sucked into the intake, chunks of dirt and sludge are going with it. So it's important to dump the old oil and wipe the oil cup out on a regular basis.

Finally, you'll recall that the air cleaner was designed to be just the right size to match engine air intake. Otherwise, too much or too little oil is drawn into the filter canister. That means that any replacement air filter needs to be similar in size and design, if not identical to the original. By the same token, if you make dramatic changes in the engine that are going to affect air intake, you will need to make comparable changes in the air-cleaning system.

When it comes to repairing an oil-bath air cleaner, your biggest enemy will likely be rust, particularly if the oil in the bottom pan has long been replaced with water. Although some restorers have had success rebuilding small holes with epoxy or J-B Weld, about the only option when you're faced with a rusted-out canister is to locate a replacement. That can be a challenge, though. Brian Holst, parts manager for the John Deere Collectors Center, explains that his staff recently discovered a stash of unused oil-bath oil cleaners in a John Deere parts warehouse. Unfortunately, over half of them were rusted beyond use.

To work correctly, the oil-bath air cleaner cup must be filled to the recommended level with the correct weight of oil.

Manifold Inspection and Repair

Just as an air leak in the carburetor can affect how smoothly your tractor runs, so can a crack in the intake manifold. When that is the case, one option is to have a local welder or machine shop make the repair. Unfortunately, most intake and exhaust manifolds are made out of one of four types of cast-iron material—white, gray, malleable, or ductile iron. To weld it, the material must be properly prepared, preheated, and welded with the appropriate method. However, the type of material must first be identified, which involves one or more of the following tests: chemical analysis, a grinding test that identifies the types of sparks a grinding wheel gives off when in contact with the material, and a ring test that helps identify the material by the type of ringing sound it gives off when struck with a hammer.

Should you decide to try arc welding a cracked manifold yourself, it's important to use a high-content nickel/cast rod or a nickel/cadmium rod with a cast-iron-friendly flux. Also, try to preheat the manifold with a torch. If you try to lay a long bead of weld on a cold manifold, it could easily warp and cause a sudden stress crack somewhere else. If preheating is not possible, strike an arc and weld only an inch or so of the crack. Then stop and let the heat spread to other parts of the material.

One alternative to arc welding is brazing the manifold using a brass rod melted into a prepared groove on the manifold crack. Start by locating the crack and grinding a groove along its length with a grinder. Extend the groove a half inch or so beyond the crack. Then use a coarse file to remove the grinder marks from the groove. This helps remove any graphite particles that could prevent the brazing material from adhering to the iron.

As for the brazing material, it's best to select a brass rod that is high in copper content with some nickel added. Also, select a torch tip that has a high heat output with low gas pressure. As with arc welding, it's helpful to preheat the material to be welded so it won't crack under isolated heat stress. Once brazing has been completed, try to cool the manifold slowly, using a bed of sand, if available.

One last option when trying to repair a manifold is to use an epoxy, such as J-B Weld. This is a particularly viable option on an intake manifold that is in a relatively cool area of the engine and when the crack is not in a stress area, such as the areas around the mounting flanges. Just make sure the area is clean, free of grease and grit, and prepared according to the directions on the epoxy package.

Of course, the easiest alternative is to just locate a replacement manifold at a salvage yard, flea market, or aftermarket vendor. If you have to hire someone to do the welding, finding a replacement may be the cheaper alternative as well.

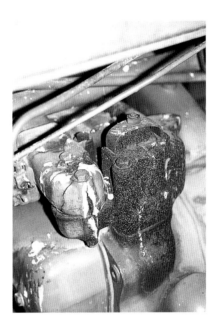

After years of exposure to the elements, the manifold material on some tractors simply cracks and disintegrates.

When an intake manifold gasket is no longer available, you have to settle for making your own.

Governor Overhaul

While it doesn't come in contact with the fuel, the governor plays an important part in the fuel-delivery system and needs to be inspected as part of your overhaul. In general, the governor uses a rotating mass applied against a spring to adjust the carburetor throttle shaft and thus regulates the engine speed around a set point established by the throttle lever position.

On virtually all horizontal two-cylinder John Deere tractors, the governor is a centrifugal flyweight type driven by the engine camshaft gear. The governor housing includes a fan drive pinion, mounted to the governor shaft, which is in constant mesh with the fan drive bevel gear. On models so equipped, an idler gear, which is mounted in the rear portion of the governor cases, also drives the live Powr-Trol pump.

In contrast, the governor on vertical two-cylinder tractors is mounted on the front of the engine, where it is linked to the carburetor and throttle lever through a series of springs, levers, and linkage rods.

The first step in governor inspection is checking for any signs of malfunction. Symptoms can include the engine idling too fast or not idling down when the throttle lever is moved to the idle position; surging; over-revving; the engine not reaching the specified top speed; engine speed control that is erratic; and delayed reaction or sluggish response to changing load conditions or throttle movement.

Before removing the governor or attempting any disassembly, inspect all linkages and link rods for free movement and the absence of any bends or binding. If necessary, free up and align all linkages to remove any binding. If additional internal repair or inspection is required, follow the instructions in your repair manual for governor removal and installation. On some early models, it will be necessary to remove the entire governor assembly, while the governor shaft and weights on later models, such as the 20 Series, can be removed without removing the housing.

In most all cases, the fan drive bevel gear mesh is adjusted with shims. Therefore, it's important you don't misplace or forget to reinstall the shims between the fan shaft rear bearing housing and the governor case during reassembly. If the bevel gears are not to be replaced, the same shims should be reused.

The basics for governor overhaul include inspecting and replacing any defective bearings, seals, and drive gears. Also, ensure that the flyweights move freely without binding.

If you're not sure of your ability to overhaul or rebuild the governor as instructed in your repair manual, you might consider sending it out to one of the shops listed in the appendix that specializes in governor restoration. Considering the role it plays in controlling engine speed, you want to make sure it's done right.

When reinstalling the governor, make sure the timing marks on the governor drive gear and camshaft gear are in register. If the marks have worn away, follow the directions in your tractor service manual, using the "L.H. Impulse" mark on the flywheel to correctly align the gears.

Finally, tune the speed adjustment as necessary to limit the engine speed to the rpm rating specified in your service manual. On A, B, G, and H models, the adjusting screw is on the steering column next to the throttle. On Model D tractors, the speed adjustment is regulated by the position of a sleeve on the throttle rod.

Above: The governor mechanism on most Waterloo-built tractors, including this Model D, also drives a shaft to the radiator fan.

Left: Be sure all the links between the governor and the carburetor are free of binding and operate freely, particularly on Dubuque-built tractors.

Below: Inspection of the governor should include making sure the flyweights move freely.

CHAPTER 14

Cooling System

For those used to today's liquid-cooling system—which normally includes a radiator, pressurized radiator cap, thermostat, water pump, and fan and fan belt (or, on the newest models, a thermostatically controlled electrically driven fan)—the cooling system on most early model tractors appears rather primitive.

Until 1952, most two-cylinder John Deere models used a simple thermo-siphon circulating system, which consisted of a radiator and two hoses that connected it to the engine block. In fact, in a 1947 sales brochure for the Model B, Deere actually extols the virtues of its simple thermo-siphon system. Because hot water is lighter, it is forced upward from the engine, through a narrowing passage, to the radiator. Here, it sinks as it is cooled, only to exit at the bottom and return to the engine.

Cooling System Inspection and Repair

You probably had an opportunity to evaluate the cooling system to some extent when you purchased the tractor or during the troubleshooting operation. Perhaps you even had a chance to start the engine and let it run long enough to see if there were any water leaks, problems with overheating, or traces of oil in the coolant.

Unfortunately, radiator cores tend to clog up with rust, lime, or other mineral deposits and the fins plug up with weeds, seeds, and debris. In addition, the metal headers often corrode away after years of use, and the seams become moist with residual antifreeze.

Hence, it's best to start cooling-system inspection and restoration at the front of the tractor, at the radiator. The first thing you should do is check the front and rear of the radiator for a buildup of bugs, seeds, weeds, and so on. A strong stream of water sprayed from the back side, or fan side, of the radiator will remove a lot of the debris.

Next, check for moisture around the radiator core and headers. These areas tend to rot out if the tractor has sat dry for a long period of time. If there is leakage, the area will be moist, and perhaps even smell sweet if there is antifreeze in the system. If the leakage is minor, you can sometimes take care of the problem by adding one or two cans of radiator "stop leak" material.

If there is substantial leakage, however, it's best to remove the radiator and have it professionally serviced. Considering that you may have to remove the radiator anyway as part of the tractor or engine restoration process, you may want to think about taking it to a professional, just to have it flushed, flow tested, and checked for integrity. The other option is to simply replace the radiator core as part of the restoration process.

One of the important steps in cooling system restoration is checking the integrity of the radiator.

Some radiators, especially those used on thermo-siphon cooling systems, can be split into three sections. This allows you to clean or even sandblast the upper and lower cast housings.

Above: This radiator core, which was designed to bolt between an upper and lower casting is definitely in need of replacement.

Above right: While restoring this Waterloo-built model, Estel Theis elected to replace the radiator core to ensure against future problems.

Right: If water leakage is not excessive, you can often take care of the problem by adding one or two cans of radiator "stop leak" material.

Shutters and Curtains

The cooling system on a large number of John Deere two-cylinder tractors was designed to do more than just cool the engine. Any all-fuel model, designed to burn both kerosene and gasoline, must have the ability to run hot on demand in order to properly vaporize the less volatile kerosene.

For this same reason, most kerosene tractors include a small gasoline tank that is used for starting the engine. Once the engine and intake manifold are hot enough, the fuel system is switched over to kerosene or fuel oil. That means the operator has to continually keep an eye on engine temperature—not just keeping it cool when necessary, but keeping it hot enough, too. To assist in this endeavor, John Deere equipped most all-fuel models with a set of vertical shutters that are opened or closed from the operator's seat. On the other hand, most models built prior to mid 1937 used a canvas curtain ahead of the radiator to restrict air flow. The curtains cannot be controlled from the operator's seat. Instead, the system utilizes a pair of clips that hold the curtain in the appropriate position after it has been unrolled and pulled down. To get the tractor warmed up quickly, the operator simply closed the shutters or pulled down the curtain and deprived the radiator of air circulation. However, there were times, such as when driving back to the farmstead from the field or when moving between fields, that it again became necessary to close the shutters to keep the engine temperature elevated and the engine running smoothly. Since most restored tractors are continuously operated on gasoline, the operation of the shutters is no longer a concern. The main role now is appearance.

Unfortunately, original temperature control devices easily suffer the effects of neglect. Over the years, dirt and chaff would naturally collect between the bottom of the shutters, the radiator, and grille. As a result, the shutters on a lot of restoration-quality tractors have since rusted out as moisture collected in the residue and did its damage. The good news is that, like many other once rare parts, curtains and shutters for most Deere models have been reproduced and are now available from a number of aftermarket vendors.

Left: **In order to raise the engine temperature to a level sufficient for kerosene ignition, all-fuel tractors were equipped with either louvers or an adjustable curtain in front of the radiator.**

Above: **Because they were exposed to dirt, debris, and moisture, temperature-control devices, such as shutters and curtains, were susceptible to rust and rot. Fortunately, replacements are available.**

Radiator Cap

There are basically two types of radiator caps used on John Deere two-cylinder tractors. The earliest type consists of a simple cap that covers the radiator opening and is held in place by a wire that snaps over the top or a knob-operated mechanism that locks the lid to the housing. Essentially, its only purpose is to keep some of the steam in and the dirt out. Due to the simplicity, about the only thing you'll need to do is clean the cap, replace the seal, if one was used, and make sure the locking mechanism works correctly. By the 1940s, John Deere followed the industry by equipping its tractors with a pressurized cooling system. Like the cooling system found in modern tractors and automobiles, it uses a radiator cap designed to raise the pressure in the cooling system so the coolant boils at a higher temperature. This, in effect, accomplishes two purposes. First, each pound of pressure raises the boiling point by approximately three degrees Fahrenheit, which allows the engine to operate at a higher temperature. Second, since there is now a greater difference between the water temperature and the air temperature, the radiator can operate more efficiently.

On systems equipped with a pressurized cap, check to make sure the bottom of the cap is clean and fits snugly into the filler neck. Check the rubber bottom for swelling, nicks, or cracks. Also check the filler neck for uniformity on the sealing surfaces. A warp or hairline crack will cause pressure to leak out when in use. Finally, make sure any replacement cap has the proper pressure relief rating. If the relief setting is too high, you run the risk of blowing hoses or the radiator core, especially if the core is weak in the first place.

Before the advent of pressurized cooling systems, the radiator was simply equipped with a cap held in place with a wire clip or rotating latch.

Fan

The fan and fan drive system on most vintage John Deere tractors are pretty basic by today's standards. There were no such things as thermostatically controlled electrical fans, and fans weren't enclosed in shrouds that helped direct air flow.

On virtually all Waterloo-built tractors, the fan is attached to the end of a long shaft that is driven by a bevel pinion gear mounted on the governor shaft. If the tractor is equipped with power steering, this same shaft also drives the power steering pump.

Whether the fan needs attention or not, you will need to remove the shaft for any type of engine or governor repair, since the shaft runs directly above the engine. While you have it off, check the integrity of the fan blades to make sure the attachment rivets are tight and the blades haven't been bent. You'll also need to check the condition of the fan friction washers that both drive and protect the fan. Because there is no fan belt, which has the potential to slip and protect the drive, John Deere devised a fan drive disk that is spring-loaded against the friction disk and fan hub. Should the fan become obstructed, the system is designed to slip the fan drive disk instead of bending blades. Excessive fan slippage is a sure tip-off, though, to the need for friction disk replacement.

Restoration also consists of checking the fan shaft bearings to make sure they're in good condition and inspecting the fan drive bevel gears in the governor housing. The fan drive bevel gears are only available as a matched pair, so if damage is evident, it will be necessary to replace both gears and adjust them for the proper mesh and backlash.

Dubuque-built tractors, on the other hand, have a cooling system that more closely resembles that of a conventional engine, due to the vertical arrangement of the cylinders. Tractors like the M, MT, and 40 utilize a belt-driven fan powered by a hub on the crankshaft. Since the hub is not fitted with bushings, restoration consists of replacing the hub, bearing, and/or shaft.

Make sure the cowling around the fan is tight and provides adequate clearance for the fan blades.

Removing the fan shaft on a Waterloo-built tractor can be a real challenge. Often you have to remove a manifold and the water-return passage.

Hoses

On many tractor restorations, the biggest problem with the cooling system—and many times the only problem—is the condition of the hoses. Hoses that are hard, brittle, or cracked need to be replaced. Keep an eye out, too, for small patches of moisture on the hose surface. If discovered, gently knead the area while looking for a hairline crack or pinhole. Such areas tend to leak only when the tractor is at operating temperature and under pressure, making them difficult to locate.

Also, look for hoses that have swelled up because of oil contamination. They feel greasy and spongy when kneaded. Replace any hoses that are marginal. While you're at it, it's a good idea to change the hose clamps, too, since dirt and grit can keep them from being sufficiently tightened to seal water.

As a final note, you'll want to consider how the tractor is to be used before replacing the hoses and clamps with conventional equipment. If you're restoring the tractor as a show model, you probably don't want to use the type of hose clamps that employ a slotted band and a worm-style tightener. More than likely, the original hoses were held in place by either the old-style spring-loaded wire clamps or wire clamps that tightened with a single screw. They may not work as well, or hold as tight, but if you're going for an authentic look, you need to go all the way.

Both photos: **Tractor restorers who are striving for originality are always in search of the old-style wire clamps that were first used to clamp hoses in place.**

Water Pump

First of all, there's a good chance the two-cylinder tractor you're restoring won't even have a water pump, since most models built during the first half of the twentieth century used siphon cooling. If your tractor is equipped with a water pump, the most likely problem you'll encounter is a water leak or worn bearings. A leaking seal is indicated by coolant leakage at the drain hole in the pump housing. Unless the impeller blades or an internal divider have been completely attacked by rust, you can usually rebuild the unit with new seals and bearings. If an impeller blade is rusted pretty badly, you may be able to find somebody to rebuild it with a welder once you get it cleaned up. Of course, salvaged water pumps aren't that difficult to find.

Note that the shaft and bearings on almost all water pump units were not designed to be serviced separately. They can only be replaced as a unit.

The first step in rebuilding the water pump is removing it from the tractor. A word of caution: Be careful you don't lose the shims located between the fan support and the tractor frame.

It's important that you inspect the water pump shaft, assuming it is being reused, to be sure it is smooth and free from rust. Otherwise, it won't be long before you're replacing seals again. If necessary, use a piece of emery cloth to smooth the shaft where it fits against the seal. A small amount of grease on the pump shaft will prevent damage to the water seal as the pump is being reassembled, especially if the impeller has to be pressed back onto the shaft.

When reassembling the water pump, make sure the highest vane on the impeller is flush with the mounting surface of the pump body. If any vane extends beyond the mounting face of the pump hosing, the protruding vane will strike the radiator's lower tank when the pump is installed.

Finally, make sure the drain hole in the bottom of the housing is kept free of dirt, grease, and paint so that any water that may leak past the seal can drain away.

Unless the impeller blades or an internal divider have been destroyed by rust, you can usually rebuild a water pump with new seals and bearings.

Above, both photos: Due to excessive rust, the pipe section of this cold water line was replaced during restoration. Note the lack of a water pump; the thermo-siphon system relied only on the laws of physics.

The water pump on most horizontal two-cylinder tractors is located at the base of the radiator and is driven directly off the fan belt.

Thermostat

Assuming your tractor is new enough to have a water pump, you'll need to make sure the thermostat is operating correctly before you finish the restoration project and put the tractor into use. Obviously, you can tell a lot about its operation by looking into the top of the radiator with the cap removed. (It should go without saying to never open the radiator cap when the engine is hot.)

As soon as the water gets hot enough to open the thermostat, you should see water start flowing into the top of the radiator from the upper radiator hose. If this is not the case, you have a couple choices. You can test the thermostat by placing it into a pan of water on a stove and watching for the diaphragm to open as the water heats up and attains the temperature at which the thermostat should open. Or, considering the price of a new thermostat and the age of the tractor, you may just want to replace it with a new one. Either way, you'll need to remove the old thermostat. On gasoline and all-fuel models, the thermostat is located in a housing bolted to the radiator water inlet casting. On LP gas models, it is located in the upper water pipe rear casting.

Belts

Because the fan on Waterloo-built Deere tractors is shaft-driven from the governor housing, a number of early two-cylinder models require no belts at all. Don't be surprised, however, if you find a generator drive pulley installed on the fan shaft, even though there is no generator on your tractor. Many tractors came with these pre-installed on the shaft; many others had a replacement shaft, which included the pulley, installed during a repair.

Even on vertical two-cylinder models and tractors equipped with a water pump or generator, the belt inventory is limited to only one or two. Still, you need to make sure any belts that are used are in good condition and not slipping.

To check a belt, twist it around in several spots so the bottom and one side are clearly visible. Look for signs of cracking; oil soaking; a hard, glazed contact surface; splitting; or fraying. Replace any belt showing these symptoms.

Make sure you adjust for the proper belt tension, as well. A belt that is too tight can cause premature wear on the bearings, while a belt that is too loose can slip, squeal, or cause other problems.

CHAPTER 15

Sheet Metal

Thanks to the huge interest in John Deere tractor restoration, there are now literally hundreds of sheet metal parts available as aftermarket reproductions. According to Brian Holst, parts manager at John Deere Collectors Center, they include complete hoods for most unstyled two-cylinder tractors and styled B models, replacement nose pieces, dash panels, grilles, and various fenders.

In addition, there are plenty of sheet metal parts available through salvage operations, such as Dennis Polk Equipment or Central Plains Tractor Parts (see appendix listings). Finally, there are a number of businesses, including many John Deere dealers, that sell what is commonly referred to as new-old stock (NOS). These are old parts that have never been used, but instead have been stored in a warehouse or stock room and have only recently been "discovered" and put into circulation.

The supply of air cleaner oil pans Holst has in stock at John Deere Collectors Center is a good example of new-old stock. They were found in a John Deere warehouse, cleaned up, and put back into circulation. Unfortunately, more than half were unusable due to rust and had to be thrown away.

A wide variety of sheet metal components, including fenders and hoods, are available from aftermarket sources.

Even replacement dash panels and replacement nose pieces are available from a number of sources.

Parts Manager Brian Holst displays some of the other replacement parts and components available to restorers through the John Deere Collectors Center.

Paint Removal

"Due to the availability of so many new parts, I personally like to replace any sheet metal parts that I can with new," says Jeff McManus, senior consultant for the John Deere Collectors Center. For many tractor restorers, however, it's more economical to repair existing sheet metal components, unless they are extremely rusted or wrinkled. Assuming the tractor and all components have been thoroughly cleaned, the first step will be to remove what paint is left. Before you do that, though, take note of any original decals that are still left on the tractor. This is another good time to take pictures. Then, grab a tape measure and a notebook, and take notes on the position of each decal.

One of the fastest and easiest ways to remove old paint, rust, and dirt is by sandblasting the engine, frame, and sheet metal. However, you need to be very careful not to damage or warp parts in the process. The fine sand that works so effectively at removing paint tends to work its way into cracks, crevices, and seals, as well. Sandblasting an engine or transmission, for example, can be very damaging if the sand is forced into the case or ruins the seals.

Consequently, before you start any sandblasting, it's important that you go over the entire tractor looking for holes that will let sand in and cause later damage. Check to see if any of the bolts you removed during disassembly left open holes that will let sand into vulnerable areas. If so, you'll want to put bolts back in these holes. Watch out for seals and drain holes, as well. The water pump, for example, has a vent on the bottom side that will let sand up into the bearing and shaft area. Abrasive sand will also damage any seal it contacts, such as the ones around the PTO shaft. Unless you plan to clean everything out and replace the seals anyway, you'll need to carefully mask all of these areas before sandblasting.

Don't count on just trying to avoid these areas, because if there is a hole, sand will get in. Scott Carlson, service manager at John Deere Collectors Center, cautions that masking tape isn't enough to protect an area, either. With enough pressure behind it, sand will go right through masking tape. Instead, Carlson relies on multiple layers of duct tape placed over any vulnerable components or openings. He says that if he doesn't remove the radiator prior to sandblasting, he also covers the radiator core with several layers of cardboard and seals the seams with several layers of duct tape.

Be sure you cover the serial number plate, as well. Depending upon the age of your tractor and the serial number, this can be a valuable component.

Wheels and cast parts can be sandblasted without much risk of damage. In fact, if you're dealing with spoke wheels, sandblasting may be the best way to get in and around the individual spokes.

When it comes to using a sandblaster to strip sheet metal, you find differing opinions. There are some restorers who sandblast all the sheet metal and every bit of the frame prior to a restoration project. And there are others who wouldn't take a sandblast nozzle anywhere near the sheet metal, regardless of how much elbow grease it saved. Carlson is one of those restorers who uses a sandblaster on everything. He insists sandblasting sheet metal is one of the quickest and most efficient methods available for removing paint and rust, providing it is done delicately—and he emphasizes the word "delicately."

For anything but the heaviest gauge steel, make sure you or the commercial operator use fine silica sand or glass beads. Keep plenty of distance between the nozzle and the steel to avoid warping or stretching the part. And always keep the nozzle moving. If the metal gets too hot, it will end up having ripples that are virtually impossible to remove.

Jim Deardorff, who cleans and paints old steel around his home in Chillicothe, Missouri, says he has discovered an even better way to sandblast sheet metal. He has developed "Classic Blast," a special blasting mix made up of aluminum oxide and ground black walnut shells. Using the product in a closed-top sandblast pot, which uses a vacuum to pull the media into the chamber, he says he can reduce the pressure to as little as 35 pounds and still clean fragile parts without damage. To prove it, he often demonstrates the effectiveness of his sandblast method by removing the paint from an aluminum pop can.

Another cleaner that is widely used by a number of restorers, particularly to remove paint from sheet metal parts, is a quality aircraft metal stripper. Just keep in mind that chemical strippers are generally toxic and require adequate ventilation. Also, you need to make sure all traces of the stripper have been removed prior to painting the part or the tractor. Otherwise, the new paint will soon strip off, too.

If you can find someone to do it for you, another good way to strip paint from sheet metal parts is to dip them in a caustic soda bath. A lye solution has the added benefit of removing any grease that may be on a part.

The last method of removing paint, and the one you're probably going to have to employ as well, is mechanical removal. Unfortunately, this method requires the most sweat and hard work. You'll find a wide range of "weapons" available at most hardware and automotive stores, but you might want to start with the basics, including wire brushes and sandpaper. Those that fit on an electric drill can also come in handy when removing paint and rust.

Sandblasting is still the best way to remove rust, dirt, and old paint from cast components like spoked wheels.

This is a good example of an engine hood that has been stripped and prepared. All that is left is priming and painting.

If you have access to a cabinet-type sandblaster that uses glass beads, even small and delicate parts like the lower half of a carburetor can be sandblasted.

If you plan to sandblast the tractor frame, don't forget to cover the serial number plate with several layers of duct tape to protect it from damage.

Just as they were with rust and paint removal, hand sanding and elbow grease are often a necessary ingredient to body work.

Repairing Sheet Metal

Once you have all the paint off, the first thing you're going to notice are all the dents, dings, scratches, and rust spots that need to be filled in, pounded out, and otherwise hidden from view. You may even have to splice in one or two pieces of sheet metal, create a new bracket, or in the worst-case scenario, fabricate a whole new sheet metal section.

Dents and Creases

Let's start with the dents and creases. If a dent or crease is more than $3/16$ to $1/4$ inch deep, it's best to smooth it out with a body repairman's hammer and dolly. Do not simply fill it in with body putty. Bondo might be fine for automotive repairs, but the vibration that is inherent with tractor operation can cause body putty to pop right out of a deep dent, leaving you with an ugly hole that will require more work and a new coat of paint.

If there aren't a lot of dents to take care of, you might be able to get by with a ball-peen hammer and a mallet or large hammer to back up the piece. The thing you have to keep in mind is that the original metal was stretched as the dent was created. Hence, you may have to shrink it as it is straightened. One way to do that is to heat the spot with an oxygen-acetylene torch before beating out the dent.

Another technique, particularly if you are trying to remove a sharp crease, is to drill a series of small holes (approximately $1/16$ inch in diameter) along the crease. This will allow the metal to shift as it goes back into place. The holes can be filled later with epoxy or plastic putty.

Repairing Holes and Rust Areas

In some cases, you may need to cut out an old piece of metal and weld in a new piece. One likely spot for repair is around the exhaust pipe where it exits through the hood. Because the muffler mounting bolts were located under the hood, a lot of original owners didn't go through the effort to remove the hood, steering shaft, and grille just to replace a muffler. They simply cut or peeled away enough of the hood to reach the bolts from the top. To them, time was more important than beauty.

Small dents in sheet metal can often be removed with a ball-peen hammer and a mallet or large hammer to back up the piece. On stubborn dents, though, it may be necessary to apply heat to help shrink the metal back into place.

If you have many dents to remove, you may want to invest in a bodyman's hammer and dolly.

Above, both photos: John Deere owners were notorious for peeling back the hood around the exhaust to save time during a muffler replacement. Cutting a hole around the bolts was a lot quicker than removing the hood, steering shaft, and everything connected to them to get to the bolts.

Left: Brian Holst displays one of the patches used to repair the hood area surrounding the exhaust on A and B model tractors.

The good news is that there is now a patch available to correct the problem on Model A and B tractors. Available through John Deere Collectors Center, among other sources, the patch is already sized and contoured to fit the back side of the hood. Plus, it has a raised section, which is the exact thickness of the hood that surrounds the exhaust pipe hole. As a result, all you have to do is cut a clean, square hole large enough to accept the raised lip and carefully weld the patch to the underside of the hood. Then, simply fill the gap between the hood and the patch and sand it smooth.

Patches to any other sheet metal area, including fenders, can be done in a similar manner. First, remove all the paint from the area to be worked, if you haven't already done so. Next, make a clean cut around the damaged area so you have removed all the bad metal and have left a clean, solid edge. It's important that you cut beyond the damage, because when you take the pieces to a welder, or do the work yourself, any thin, pitted surfaces will self-destruct. You'll also want to remove the section in a shape that will be easy to reproduce. As with the muffler patch, a square section with clean right angles works best.

Now, find a scrap piece of sheet metal that is the same thickness or gauge as the original piece. The biggest mistake people make at this stage is using a slightly thicker or thinner replacement piece. You may also want to trace the hole you have created onto a piece of cardboard and make yourself a template. This will be particularly helpful when you go to cut out the new piece.

If necessary, bend the new piece to match the contour of the hood, grille, or fender where it is being installed. Finally, clamp or tape the new piece into position and tack weld it in place. Then, finish welding around the splice, being careful not to heat the area to the point it warps or disfigures the sheet metal. You'll want to hide as much of the weld as possible, too. Some restorers simply use a series of spot welds around the patch and fill in the seam with body putty a little later. Throughout the process, be careful not to set the welder at too high of a temperature and burn through the sheet metal you're trying to repair.

To complete the patch, grind the welds down to remove any high spots and fill the area with J-B Weld, body filler, or Bondo. This will also fill any rust pits and gaps that have been left. Once the filler has hardened, you can sand it down to the point where the patch is flush with the clean sheet metal, using finer and finer sandpaper to finish the surface.

If you're dealing with rather small holes, you can often get by with putting a piece of fiberglass cloth on the back side of the cavity and filling it with Bondo or body filler. It may take several thin coats before you get the hole filled to surface level. At that point, you can sand and treat it just as you would any other patch.

There may be times when it is necessary to splice in a new piece of metal onto a rusted-out area. Spot-weld a metal piece of comparable thickness in place and then cover the patch with body filler. Sand the entire area to a thin layer that only covers imperfections.

With the paint removed, this hood piece is ready to be finished off with a little body filler.

Your local auto-parts supplier can direct you to everything you'll need to fill and cover minor imperfections.

A thin coat of body filler is ideal for filling and covering all the pits left by rust. However, it should not be used to correct deep dents and creases.

Once any added body filler has hardened, it should be sanded to a smooth finish that is flush with the original sheet metal surface.

Using long sanding boards, whether you're sanding by hand or with a power sander, will help ensure a smooth surface when finishing body filler or a coat of primer.

Don't be surprised if large sheet metal parts like fenders and hoods don't fit perfectly when it comes time to reassemble the tractor. It doesn't take much to shift a bolt hole ever so slightly—particularly if you had to apply heat to a part.

Because of its somewhat fragile nature, the corrugated screen used in many tractor grills is often torn, dented, or gouged. Fortunately, replacement material is available. A new grille screen makes a tremendous difference in the appearance of this restoration.

CHAPTER 16

Paint

Regardless of how good of a job you did on restoration and repairs, the first thing people are going to notice on your restored tractor is the paint job. Small imperfections may not even be visible at this point, but rest assured, once they're covered with a coat of paint, they will show up like neon lights. Consequently, it's important to take your time and do it right. The first step is to lay down a foundation of primer, followed by multiple coats of enamel or lacquer.

The Primer Coat

Once the body and frame have been cleaned and prepared, and the sheet metal has been stripped, smoothed, and filled as necessary, it's time to apply a quality coat of primer. Some restorers recommend applying a coat of primer immediately after the tractor frame has been cleaned, even if it won't be painted for a while. While the primer isn't guaranteed to prevent rust, it does make it easier to clean away any grease or oil that shows up.

That brings up another tip. Most experienced restorers insist it is best to start up the tractor and drive it around a little, if possible, before applying a coat of paint. You don't have to have the sheet metal or fenders back on yet. The goal is simply to find any leaks now, and to discover unresolved problems in the transmission or engine before that shiny coat of paint has been applied.

The primer stage also gives you the opportunity to take care of a lot of the imperfections that may remain after most of the body work has been completed. By putting on two or three coats of primer and sanding between each coat, you can easily fill a lot of pits and crevices.

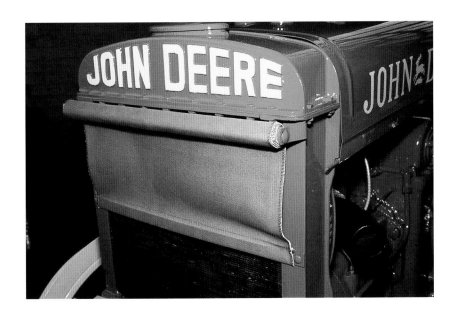

A carefully applied coat of paint really makes the raised letters on the front of this Model D stand out.

Filler Primers

For even more complete coverage of imperfections, you might want to follow the lead of other tractor restorers and use a filler primer, or heavy-bodied primer, which fills in pits even quicker. You may even want to use a little spot putty to eliminate any air pockets and holes found in the filler primer after it dries. Between coats of primer, be sure to use a fine-grit sandpaper, such as a #240 grit, to smooth the surface. On the last sanding before painting, you'll want to use an even finer sheet, like #400 grit, if you're using enamel and up to #600 for urethane or lacquer.

Jeff McManus, senior consultant for the John Deere Collectors Center, says he may apply up to fifteen coats of filler primer on a sheet metal part that has had a lot of work, just to get a glass-smooth surface. Between every three coats, he uses long sanding boards to smooth the finish. This helps prevent high and low spots from being formed by the sander itself. In the meantime, he often switches between different colors of compatible primers so that he can better identify hills and valleys in the finish. He finishes with finer and finer grades of sandpaper, ultimately ending up with wet sanding paper.

Although many so-called filler primers are actually lacquer-based primers, some urethane primers are also called filler primers. One of the beneficial characteristics of a urethane filler primer is the fact that when you sand it, the material will actually reflow and close up. Urethane primers are usually 1 to 2 millimeters thick per pass, says Hermie Bentrup, a paint specialist with Auto Body Color in St. Joseph, Missouri. So if you go around the tractor three times, you've got 6 millimeters of primer on there, whereas most new cars have about 7 millimeters total, including all primers and paints. In the end, you'll have a thicker, tougher finish that will last for years.

It's important to note, however, that any urethane primer will need a sealer coat before the top coat of color is applied.

Sealing Primers

The final step before applying a coat of paint should involve applying a coat of sealing primer. This closes the surface and prepares it to accept a coat of paint. Before you apply the sealer, though, use a tack cloth to get the surface extra clean. You don't want to seal in any dirt or sanding dust.

Choosing the Right Primer Type

There are a number of different types of primers that can be used to prepare, fill, or seal the surface prior to the final coat of paint. Each has its own unique role and application. For example, while epoxy primers protect components from new rust, urethane primers tend to form a harder finish. However, you need to ensure that the primer you select is not only compatible with any paint that remains on the tractor or engine, but also with the paint you have selected to finish the project.

Bentrup explains that the type of primer you start with depends to some extent on the type of finish you're covering—old paint, bare metal, or cast iron.

Epoxy Primers

If you're shooting primer over bare metal or cast iron that could be exposed to the weather before you get it covered with additional primer, Bentrup generally recommends an epoxy or self-etch-type primer. This is particularly the case with parts of the frame or cast wheels that have been sandblasted and are in no need of further work or sanding.

Epoxy is the easiest to use because it combines the qualities of a metal etch, a primer surfacer, and a primer sealer in one product. A self-etching primer, on the other hand, is basically a phosphoric-acid-type etch.

Self-etching primers have to have another primer over the top of them, Bentrup adds. It can't be an epoxy primer, but it can be a urethane primer. You cannot paint directly onto a self-etching primer, because the paint won't adhere. That's why painters generally recommend that tractor restorers go over bare finishes with an epoxy primer, since it will give you a more durable finish, and it will etch aluminum and metal in one shot.

Epoxy primer can be used in one of two ways. It can be used as a primer-sealer, where you spray it on, wait 15 to 20 minutes, and start top coating with your color. Or it can be used as a primer-surfacer to cover minor flaws in the surface. In this case, you will want to put down two to three coats, giving it 15 to 20 minutes between coats. Then, wait at least 6 hours before sanding the surface.

Finally, epoxy primer can be sprayed over the top of old lacquer paint—which was often used on older tractors—without a problem. It is a good idea, however, to seal the original lacquer just to be sure the two surfaces remain compatible. It is not recommended that you ever put lacquer on top of enamel.

Urethane Primers

Although urethane primers are very popular with restorers due to their hard finish, they do not include any kind of chemical agent to prevent rusting. Therefore, you either need to ensure that the surface is completely free of rust before you apply a urethane primer,

You'll find several different kinds of primers on the market, including filler primers and those that are self-etching. Your paint supplier can help you select the best type for your needs and application.

or you need to lay down a coat of epoxy primer or an etching primer and put the urethane over it. Otherwise, you may find rust popping through the surface four or five months down the road.

Since urethane is only a primer-surfacer, you'll also need to apply a sealer of some kind before you paint. Again, a coat of epoxy primer over the top will serve the purpose. Plus, a coat of epoxy will serve to bridge any scratches in the urethane that have been left after sanding and leave a smooth surface for the top coat color.

The only thing you shouldn't do with urethane is put it under a top color coat of lacquer.

The Color Coat

For John Deere enthusiasts who are ready to paint their labor of love, the good news is John Deere tractors have always been green and yellow (unless you're restoring an industrial model). That wasn't the case with nearly every other tractor manufacturer, as they changed from shades of gray or blue to something as bright as orange or yellow. Even then, the color shades changed over the years.

So does that mean you can just pick up any can of "John Deere Green" paint and get yourself a paint sprayer? Not necessarily. First of all, don't assume you can go to the implement store or a farm supply and get the correct paint color. Granted, the farm and ranch supply outlet may have cans of paint labeled "John Deere Green." However, they may or may not be an exact match to the paint color originally used on your tractor. Canned paint can also vary in color from one can to the next. Hence, these cans are generally best left to farmers who want to touch up their work tractors.

If you're going for a true restoration, you have a couple of choices. One is to go to an automotive paint supplier and have the paint custom mixed to the true color. The other is to go to the John Deere dealership for your supplies. Keep in mind, though, that John Deere changed the shade of green around the time it introduced its New Generation tractors. If you shop for paint at the dealership, you'll find they have "Agricultural Green" and "Classic Green." The latter is primarily stocked for two-cylinder models, although there are a number of John Deere restorers who prefer the Agricultural Green used on modern tractors. The choice is really yours.

According to technical representatives for the Two Cylinder Club, nearly 80 percent of the tractors at John Deere Expo events have John Deere paint. On the other hand, Dan Peterman, with Rusty Acres Restoration, says he prefers a regular acrylic enamel, specifically a product produced by Martin Senour.

Selecting the Right Paint Type

Just as you did with the primer, you'll need to decide what kind of paint you want to use on the tractor. While some restorers prefer acrylic enamel, others opt for lacquer or the new polyurethane finishes.

Lacquer Paints

One of the things that made lacquer the choice of previous generations is the reason you seldom see it used anymore—that is its vaporization and evaporation characteristics. Lacquers tend to dry very quickly because of the rapid evaporation of the solvent used as a carrier. For that reason, the EPA would prefer nobody used them anymore.

According to paint specialist Hermie Bentrup, lacquer has other disadvantages, too, including the fact that it dries to a dull finish and must be buffed to bring out a shine. Lacquers are also the most photochemically reactive, which means they fade over time when exposed to sunlight—though newer acrylic lacquers offer improved ultraviolet radiation protection. Finally, lacquers do not withstand exposure to fuel spills and chemicals, as well as other types of finishes.

On the other hand, lacquers are probably one of the easiest of the paint types to apply, Bentrup says. If you make a mistake, you just sand it down and paint right back over it. With enamels and urethanes, there are steps you have to take to recoat it.

Acrylic Enamel Paints

Perhaps the most popular paint type these days is enamel, since it is available in a broad range of colors, allowing it to be custom mixed to match virtually any tractor color. In addition, enamel is relatively forgiving and requires minimal surface preparation. The downside is it takes a little longer to dry and must be applied in multiple, light coats to keep it from running.

Another characteristic of acrylic enamel is that it dries from the outside in. This means that the underneath side of the coat is still porous for quite some time. As a result, if you spray back over it too soon, the new coat will work its way under the top layer and cause it to lift or wrinkle.

For that reason, Bentrup recommends the use of a hardener, which causes the coat to dry from the inside out and allows recoats without a lift problem—the next coat can be applied as soon as the first coat is dry to the touch. In addition, an acrylic hardener will increase the gloss and provide a more durable finish. Adding a hardener has the negative affect, though, of reducing the pot life of the mixed paint. In the case of enamel, the pot life goes from about 24 hours without hardener to about 4 hours once the hardener has been added.

Urethane Paints

Urethane paints, which are actually part of the enamel family, are becoming popular. Among the reasons are the fast drying time compared to acrylic enamel, and the durability and luster that accompanies the hard finish. On the other hand, urethane paints are not available in nearly as many colors as acrylic enamels.

Bentrup notes that there are basically two types of urethane in use. One is a base coat with a clear-coat finish, which is what most of the automotive manufacturers have gone to. In essence, it's a cheaper route, even though the car companies would have you believe it is a superior finish. In effect, less pigment is used to lay down a color, and the gloss comes from the clear coats that go over it. As a result, base-coat finishes require at least two to three coats of clear coat for both protection and shine.

The other type of urethane is a single-stage urethane. Like acrylic enamels, there are some single-stage urethanes that don't need to be clear coated, although both single-stage urethanes and acrylic enamels can be clear coated for additional shine and protection. In the absence of a clear coat, two or three coats of single-stage urethane are recommended.

Ensuring Paint Compatibility

Although restorers all have their favorite choices of paint type, it's important to ensure that the primer, paint, thinner, and clear-coat protectant are all compatible with each other.

One tractor restorer experienced the frustration of having to repaint a hood and grille three times because the paint bubbled as it dried. After several attempts to figure out the problem, he finally traced it to the hardener, which was either bad or incompatible with the paint he was using.

"We recommend that you stay within the brand that you are shooting to ensure compatibility," says Bentrup. "Just like everything else, there are off-brands that will work. But we try not to recommend them, because if there is a problem, then you've got everybody pointing fingers. If you're within the brand, you know everything has been tested in the lab for compatibility and, unless you've made a mistake, the company has to stand behind it."

It's important that all the paint products you are using—including primers, paint, thinners, and hardeners—are compatible.

Painting Equipment

Selecting the right paint for your tractor is one thing. Gathering the right equipment to protect yourself while completing the task is quite another. In the long run, the latter is really the most important At the least, protective equipment should include rubber gloves, protective clothing, and an approved respirator or mask.

To be safe, use a respirator approved for organic mists, which is the type labeled for use with pesticides. While a charcoal-filter mask may be sufficient for enamel paint, urethane coatings and acrylic enamels to which a hardener has been added contain chemicals known as isocyanates, which are especially toxic. Hence, the use of urethane requires a fresh-air mask and painting suit, or a charcoal mask and a fully ventilated environment.

Remember, too, that paint, thinners, and solvents are highly flammable, particularly when atomized by an air-powered spray gun. So be sure the area is free of any open sources of ignition and keep a fire extinguisher nearby.

Finally, make sure your air compressor and paint gun are adequate for the job. Most restorers use a siphon-type gun that siphons the paint out of a canister or cup and draws it into the air stream. If you want to spend about twice the price for a new gun, you can move up to an HVLP unit, which stands for high-volume, low-pressure painting. This type of gun feeds the paint directly into the air stream, which tends to save paint and generate fewer vapors.

Regardless of what type of paint gun you use, make sure your compressor will provide enough air capacity and that you have enough hose to move freely around the tractor.

A respirator approved for use with organic mists is one of the most important pieces of equipment when painting a tractor. Paint fumes are not only hazardous to your health, but they are also flammable.

One of the requirements for a good paint job is a quality paint gun and a compressor with adequate capacity. One of the newest types of applicators is high-volume, low-pressure gun (HVLP), which helps conserve paint and reduce vaporization.

Applying the Paint

If you're like most restorers, you can't wait to get the tractor assembled and get it painted. But try to restrain yourself just a little longer. You'll achieve the best results and have the easiest time painting your tractor if it is still disassembled. That means you should look at painting individual sections of sheet metal, as well as the frame and engine, separately, whenever possible. Some restorers even prefer to paint the wheels with the tires removed, rather than masking them, to avoid the potential for overspray on the rubber. Just make sure the paint has had plenty of time to cure before remounting the tires and be ready to touch up any blemishes.

By leaving as many parts off the tractor as possible, you also have the opportunity to paint both sides of a piece in one session. Components like the seat, grille, battery cover, and so on can be hung on wire hooks, for example, and coated on all surfaces, without having to let one side dry first. If you're painting a two-tone tractor, like the 20 or 30 Series, most restorers suggest masking the tractor and painting the yellow areas first.

It's important to adjust both the spray mixture and the spray pattern. One, of course, can affect the other. In general, several thin coats of paint are better than one or two thick ones. On the other hand, if you get the paint too thin, it can have a dusty appearance that reduces gloss and shine. To attain the right consistency, you'll need to add thinner or reducer. According to Bentrup, these two components do basically the same thing: They improve the spray pattern and the paint's ability to evenly coat the surface. It's just that they're usually referred to as thinners when used with lacquers and reducers when used with enamels, including urethane products.

Once you have attained a mixture that sprays smoothly and evenly, adjust the nozzle to spray an oval that is approximately 3 inches by 6 inches at about 1 foot distance. When painting large areas, spray around the edges first and finish up by filling in the center. Concentrate on moving the gun in a back-and-forth motion to produce a smooth, even coat.

As with many things in life, practice makes perfect. So, if you don't have much experience with painting, start by practicing on a few scrap pieces of metal.

Most restorers and paint suppliers recommend at least two to three coats of paint; although some use up to five or six on sheet metal for extra durability and shine. While it may sound like overkill to some, Jeff McManus says he puts at least seven coats of John Deere brand enamel on sheet metal, putting one coat on right after the other. After he finishes the last coat, he lets it cure for three or four days, then wet sands all sheet metal with #1200-grit sandpaper, staying clear of any edges. Finally, he applies the decals and sprays clear coat over both the paint and the decals to produce a gleaming shine. Any cast parts, however, receive only paint to avoid a glossy, unnatural appearance.

Bentrup notes that one way to ensure adequate paint coverage is to stick a special black-and-white check panel on a masked area before you start painting. Once the black-and-white grids on the check panel are completely covered and hidden, you know you have sufficient coverage; any additional coats are simply building up the finish.

Finally, make sure the timing between coats is sufficient, particularly if you're using an enamel without a hardener (see the explanation about enamels).

On the other hand, since the coats rely on a chemical bond for adhesion, you don't want to wait too long between coats either. Most paint coatings, or your paint supplier, will provide some kind of guidance, so follow the recommendations.

Tractor wheels are best painted with the tires removed. Estel Theis usually places them between two secured sawhorses, which allows him to paint both the inside and outside surfaces.

Hanging small parts on coat hangers or pieces of wire allows you to paint all sides of the part in one pass.

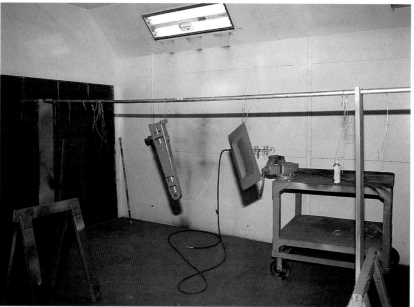

One trick that saves a lot of hand painting is pushing fasteners into a piece of cardboard or inserting them into holes drilled in a board. This lets you spray paint all the bolt heads of a like color in one pass.

When painting a tractor with two colors, such as later model 20 or 30 Series tractors, it's vital that you carefully mask off each colored area. Most restorers also recommend that you paint the yellow areas first.

Left: When painting a tractor, it's important that you use a thinner or reducer to obtain the correct consistency. Several thin coats are better than one or two thick ones. Adding a hardener will improve drying time and strengthen the paint coat.

Below: The frame and engine are best painted separately. The same goes for sheet metal parts. The tractor can then be reassembled once all components have dried and cured.

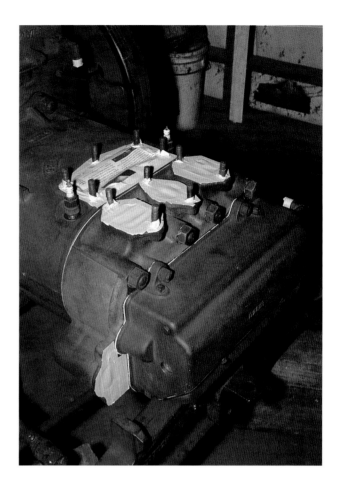

Be sure to carefully cover all openings and items such as the serial number plate with masking tape before painting the engine and frame.

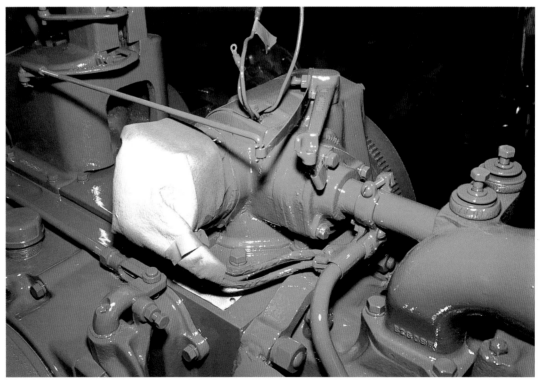

If components like the magneto/distributor and dash panel have already been reinstalled, be sure to mask them off to protect them from paint overspray.

Painting a tractor is much easier if you have room to spread out the parts and paint them individually before trying to assemble them into a finished project.

With all painted parts and sheet metal back in place, this Model 620 is ready for decals.

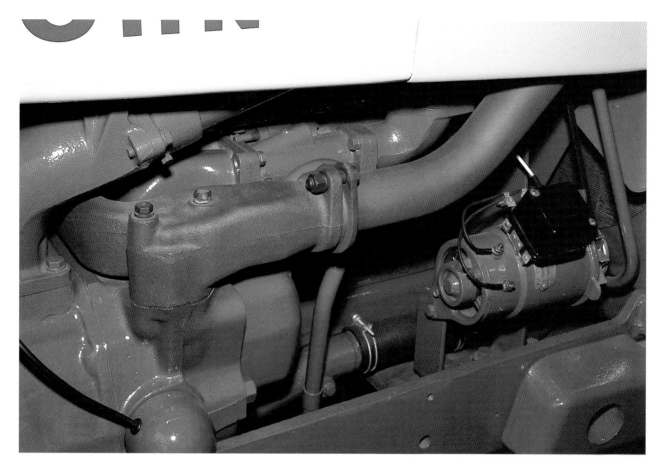

Jeff McManus, who has restored more than seventy tractors for himself and Deere & Company, likes to use a "cast gray" pigment on manifolds and rear axles.

Finishing the Axles

Traditionally, the rear axles on John Deere tractors have been left unpainted at the factory to prevent any problems with tread-width adjustment. However, that has left a number of restorers wondering how to treat the axles in a restoration project. Without a coat of paint or lubricant, unprotected axles can quickly develop a light coating of rust that detracts from the whole. Yet few people want to subject tractor show onlookers to a coat of grease and the prospect of soiled clothing.

In most cases, flat black paint has been the preferred choice, although there are others who opt for gloss or semi-gloss on adjustable-tread models. Technical advisors with the Two Cylinder Club say they've seen some John Deere enthusiasts use another option: spraying clear coat on a rear axle that has been thoroughly cleaned and prepped. Clear coat comes in spray cans, making the job even simpler and easier.

Jeff McManus says his choice for rear axles is a "cast gray" pigment that comes in aerosol cans. Because it looks almost identical to steel, he also applies it to the PTO shaft and the manifolds.

CHAPTER 17

Decals, Name Plates, and Serial Number Plates

When it comes to decals, the good news is they have never been easier to apply or easier to find. The interest in antique tractor restoration—coupled with advances in computer-graphics technology—has seen to that.

The reproduction decal business started when a few collectors commissioned a printer to produce a set of decals that they couldn't locate. Soon they found there was a demand for the product and began selling the decals to other collectors. Using modern scanning technology, decal manufacturers are now able to produce decals from drawings, literature, operating manuals, or even pencil rubbings.

The other thing that has changed is the type of decals available. At one time, almost all of the decals on the market were the old water-transfer type, similar to the ones we used to apply on plastic airplane models. After soaking them in water, the backing paper was slipped off, and the decal was applied and left to dry.

The history of decals at Deere & Company goes back to 1941, when the company started applying water-transfer decals on which the letters and numbers were screen printed. Prior to that, the company name and model numbers had been screen printed directly onto the metal. Unfortunately, water-transfer decals have a limited shelf life, since they tend to crack easily. They also begin to weather and crack very quickly after they have been applied to the tractor. As a result, water-type decals are seldom used today, especially since technology has helped decal suppliers to develop better alternatives.

These days, many of the decals sold and used are made by screen printing onto a variety of materials, including Mylar plastic. With this type of decal, you simply remove the backing paper, which protects the adhesive and carefully press the decal into position, much like you would apply a bumper sticker.

Equally popular are vinyl-cut decals. Unlike the mylar decals, the decal is sandwiched between two layers of protective paper. However, each letter or number is individually cut out of vinyl. Hence, the paper on the back protects the adhesive, while the paper on the front holds each of the letters in place as they're being applied. Unfortunately, vinyl-cut decals are often a one-shot deal, without any chance for adjustment, since the letters are all separated from each other.

Researching Decal Originality and Placement

Before you get started with any type of decal, it's important to have the right tools and the right information. The latter is particularly important. You may think you know where all the decals go, but even experienced restorers get fooled at times. Decal configurations occasionally changed from one year to another, even within the same model. And not all models were equal, either.

For example, the "John Deere" logo decal is centered on the side of the hood in the majority of models that were built through the early part of the century. However, according to Travis Jorde, owner of Jorde's Decals and an authority on John Deere decals, there were a number of variables over the years. As an example, the words "General Purpose" appeared on the side of the A and B along with the words "John Deere" until the middle of the 1935 production year, although it continued to be used on the other models, including Model G, for a little longer. At the same time, Deere dropped the leaping deer that had previously appeared between the words "John" and "Deere."

In 1941, Deere changed the lettering style to outline the words "John Deere" in black. However, there was another lettering style in 1942, leaving the 1941 year in a class by itself.

Meanwhile, the model designation was first applied on the rear of the fuel tank until 1935, at which point it was moved to the seat support, where it remained through 1946. The following year, Deere moved it to the hood where it was displayed in a round black circle.

Interestingly, Jorde notes that the seat position decals always included designations for any sub-models, such as AN (narrow for single front wheel), AW (wide front axle), etc. However, after the company introduced the round decal, only the orchard and wheatland versions of the A (AO and AR) were identified. An A with a single front wheel was just an A. The seat support "designation" decal was always positioned so that it was readable from the pulley side.

Another noteworthy highlight in the decal progression occurred in 1949, when Deere moved the "John Deere" decal toward the front of the hood. An exception was the Model G, which had a shutter bracket that was riveted in place in a location where the rivets interfered with a forward decal position. Hence, the decal on the G remained centered until the shutter bracket was welded in place behind the hood. At that point, it, too, received the forward-mounted decal.

Experienced John Deere restorers insist you need to be careful when buying a decal set that includes logos and lettering for more than one tractor model. If there are extra decals in the set, people often feel like they need to use them.

As an example, there's one decal that comes in a number of sets that states, "Keep valve closed when not running." Unfortunately, people often think that if it refers to valves it must go somewhere on the engine. As a result, you'll see that decal stuck in a variety of places. The restorer figures if it's in the kit, it must go somewhere. The fact is, the decal refers to the valve on the LP tank and is applied to the side of the tank on LP tractors only.

To help avoid such confusion, Jorde markets kits for most two-cylinder John Deere tractors that include only the decals that are to be used on that particular model. Of course, that means you need to know the year and model number. In cases where there was a mid year change in decals, you'll also need the serial number if you want to ensure accuracy.

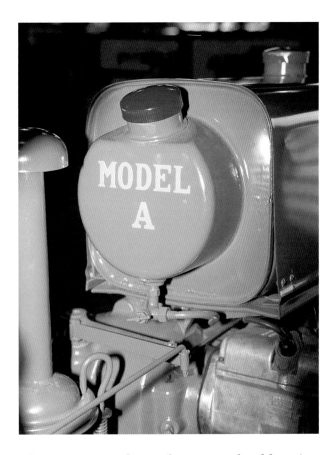

It's important to know the correct decal location before you start this final stage. In the case of early A and B tractors, the model designation was on the back of the fuel tank.

In 1949, the "John Deere" decal was moved toward the front of the hood on most models. The Model G was an exception. The decal wasn't moved until the 1950 model, when the shutter bracket was welded to the inside of the hood and the rivets eliminated.

Unless you've taken measurements before you stripped off the old paint and decals, or used an accurately restored model as a reference, it can be difficult to position decals in the correct location, especially if they don't run parallel to a seam or the edge of the hood.

The model number decal that fits on the seat bracket always includes designations for the submodel. It should always be positioned so that it is readable from the belt pulley side.

Above left: The leaping deer, positioned between the words "John" and "Deere," was used until 1935. The words "General Purpose" were also found on many early decals.

Above right: Don't forget about the small instructional decals that were a vital part of the original model. However, it's important to know which ones were used and where, since decal kits often apply to several tractor models and include extra decals.

Remember that some decal sets include decals that are not appropriate for your tractor. One example is the "power steering" decal above. Another is the one for LP tractors that states, "Keep valve closed when not running." Use only the ones that are accurate.

Tools and Supplies

In one respect, applying decals is no different than overhauling the engine—you need the right tools and supplies before you start. These should include a roll of paper towels; a clean, soft cotton towel; a roll of masking tape or drafting tape; and a rubber or plastic squeegee (you can find these in most craft, automotive, and wallpaper supply stores). A sponge may be helpful, as well. Plus, you might want to have a pair of tweezers handy for holding the edge of smaller decals. If you're using Mylar decals, you'll also need a water tray for wetting the decals before they are applied.

Surface Preparation

The surface needs to be thoroughly dry before applying decals. If you are applying Mylar decals to any painted surfaces, you also need to be sure the paint has cured. Depending on the climate in which you live, this could be anywhere from a week to a month after the tractor has been painted. If a hardener was used in the paint, you may need to wait even longer to make sure the paint isn't going to give off solvent vapor bubbles under the decal. Unlike vinyl decals, which have the ability to "breathe," Mylar is impermeable to air and vapor bubbles. So any bubbles that form under the decal after application will stay there. The paint surface must be smooth and absolutely clean, as well. If there are any pits or surface imperfections, the decal may not adhere properly. Finally, make sure your hands are clean.

You need to make sure the room temperature is within a comfortable range, too. Decals don't do well when the air temperature or metal is too cold. If you're applying vinyl decals on a hot surface or under a bright sun, they can get too warm and stretch as you're trying to stick them in place and smooth them out. The shop should be between around 60 and 90 degrees Fahrenheit to ensure that the decals adhere correctly.

Decal Application

Now that everything is ready, the first step is to hold the decal in the proper location and tack it in place or mark the edges with a few pieces of tape. You should be able to see the actual decal outline, even if it does have a protective film on each side. At this point, it's simply a matter of peeling off the backing and applying the numbers or lettering to the tractor even though the application techniques differ dramatically between Mylar and die-cut vinyl. Still, there are a few tips that will make the job easier.

Mylar Decals

If you're using Mylar decals, you might do like many veteran restorers and start by making sure your hands and tools are clean and wet. This will help keep the decal from sticking to surfaces it's not supposed to.

Some researchers like to use a spray bottle filled with water and a single drop of soap to spray the metal surface. Next put some water and another drop of soap in a cake pan or similar-sized container, and run the decal through the pan of water before placing it on the metal.

Keep in mind that there's a difference between a drop and a squirt. You only need a little soap to help break the surface tension of the water.

Although some restorers use Windex for the same purpose, Jorde cautions against using a formula that contains ammonia, as it can damage the paint and cause it to fade in sunlight. Some people think the water or liquid is used so you can remove the decal or shift it around if you make a mistake. However, the main reason you should wet a Mylar decal is so you can more easily squeeze out all the air bubbles. Remember, Mylar doesn't breathe, so if there are any air bubbles under the decal once it dries, they're really tough to get out.

Once the decal is in the exact location you want it, use the squeegee to press the decal into place and remove any water and air bubbles from beneath it. Start in the center and work outward. Then, use a soft cloth to dry the surface and remove any adhesive left on the surface.

Decals, Name Plates, and Serial Number Plates / 203

The first step in decal application is positioning the decal in the proper location and tacking it in place with pieces of tape.

Next, run a piece of tape across the top of the decal to act as a hinge. Cutting the "John Deere" decal between the words helps make it more manageable.

By lifting the decal on its tape hinge, you can then peel off the paper backing.

Then drop the decal directly into position and press it down firmly.

Once the decal has been pressed in place, carefully peel off the protective paper that holds the individual letters or numbers in place.

As a final step, wipe the surface with a soft, clean cloth to dry it and press out any air bubbles.

Vinyl Decals

Unlike Mylar decals, vinyl decals can be applied almost immediately after the paint dries, since the material will allow air and solvents to escape through it. Keep in mind, though, that some decal kits will contain both vinyl and Mylar decals. So you will have to treat the latter accordingly.

When it comes to vinyl decals, Jorde has his own tricks. Some restorers simply mark the position with a few pieces of tape, peel off the backing, and stick the decals in position using the tape as a guide. Jorde's method, however, saves time and reduces the margin of error. To begin, he places the decal in the correct position and "tacks" it in place with two pieces of tape at the top. Once he has ensured it is in the right spot, he runs a piece of tape across the full length of the top. This piece of tape acts as a hinge for the decal. Now, all he has to do is lift up the decal, pull off the backing and drop it back down into position. If he is working with a long decal, such as the "John Deere" decal that goes along the side of the hood, he will generally cut it into two pieces separating it between "John" and "Deere" so it's easier to work with. To finish the job, he simply presses the decal in place, removes the top protective paper and smoothes everything with a soft, dry cloth.

Don't worry if you have a few little bubbles this time. All you have to do is set the tractor out in the sun. The pores in the vinyl will allow the air to permeate through the decal, leaving a smooth surface. As stated earlier, don't try to apply the vinyl decals in the hot sun. They may stretch.

Clear Coat or Not?

Although some restorers like to finish off the decals with a shot of clear coat, others say they never put paint or clear coat over any kind of decal. For one thing, you have to know that the decal can take it and that the protectant won't cause it to lift off the surface. Water transfer decals, for one, can't take it.

Some decals do have a tendency to yellow when covered with clear coat, even if the surface is better protected. Body shop owner B. J. Rosmolen says he never puts clear coat over any kind of decal or appliqué, including pin stripes on an automobile, simply because it's a lot easier to replace a decal in the future than it is to restore the paint finish.

Jorde agrees, noting, "If you happen to have problems with a decal, say something gouges a letter, you can replace a single letter. But if you covered the decal with clear coat, that gloss is not going to be on the new letter. Plus, you're going to have a more difficult time getting it off."

Moreover, the companies that supply the vinyl for most decals won't warranty the product if it's sprayed with clear coat; and the clear coat manufacturers don't recommend it either. Still, a number of restorers have applied clear coat over decals without any problems. Among them is Jeff McManus, senior consultant for the John Deere Collectors Center, which restores all the tractors for Deere & Company. McManus says he routinely finishes a tractor by applying the decals and giving all the sheet metal a final layer of clear coat.

In the end, it appears the choice is up to you as the restorer and your willingness to take chances in the interest of appearance.

Emblems and Name Plates

As with most early tractors, a decal was the extent of any adornment or identification on early John Deere models. However, just as automakers started adding chrome and special emblems, so did tractor manufacturers. In the case of John Deere, with the introduction of the 20 Series in 1956, the company started installing a distinctive green and yellow emblem on the front of the hood.

Unfortunately, the emblem was in a position where it could be easily damaged while the owner worked with a loader, or if he "bumped" into something in the back of the machine shed. The good news is reproduction emblems are readily available, which means it's easier to find a new emblem than to restore the original. It will look a lot better in the end, too.

Serial Number Plates

For most vintage tractor restorers, restoration of the serial number plate is not only the last step in the project, but also a source of pride. Having a tractor with a low serial number is kind of like acquiring a limited-edition painting with a low number. Consequently, tractor collectors don't take this step lightly.

If you're working with an older-model tractor, you may be lucky enough to have a brass serial number plate. If so, a good polishing with brass cleaner will suffice.

Most of the serial number plates on later-model tractors, though, were made of aluminum. Still, you can make the numbers and letters stand out in a couple of ways. If the numbers are stamped into the plate, you may just want to clean and polish the plate with steel wool or a quality cleaner.

If the serial number is raised, however, you may want to follow the lead of Dennis Funk, a John Deere enthusiast from Hillsboro, Kansas. Funk likes to paint the serial number plate with black paint. Then, once it has dried, he lightly sands the raised areas with very fine-grit sandpaper so the letters and numbers stand out in stark contrast to the black background.

The best paint and decal job in the world will lack something if the emblem that adorns the grille is not restored or replaced.

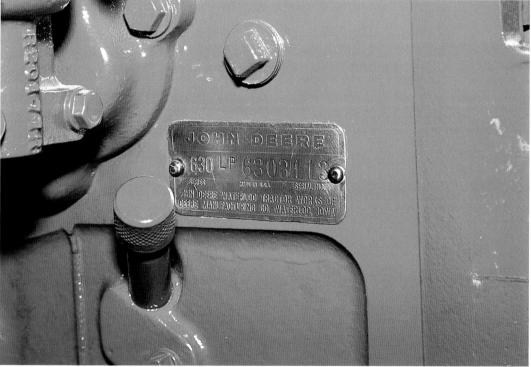

To finish off your restoration, be sure to clean and polish the serial number plate. Some restorers like to paint aluminum plates with black paint and then lightly sand the paint off the raised lettering. Brass plates are best finished with a good brass cleaner and polish.

CHAPTER 18

The Fruits of Your Labor

Although there are a number of tractor restorers who limit their hobby to restoring a desirable model and adding it to their collection, others see restoration as only one step in the pastime. For the former group, collecting antique tractors is not much different than collecting coins or stamps—just a little more expensive. The latter group sees the finished tractor as an avenue to get more involved in the growing number of clubs, tractor shows, and antique tractor events that continue to sprout up in farm country and urban areas alike.

Hopefully, the process up to this point—that of turning a pile of greasy and rusty iron into a showpiece—has been fulfilling on its own. However, there's nothing like the friendships that can be built when you get involved with a group of collectors that share your interests, frustrations, and challenges.

Above: There's nothing quite as fulfilling as that first drive on a tractor you've spent several months restoring. If you don't believe it, just ask Ed Hoyt, of St. Joseph, Missouri, who spent nearly a year restoring this John Deere Model D.

Left: Whether it's displayed at the local fair or a national show, a well-restored tractor can be a source of pride for its owner.

Antique Tractor Shows

Each year, tractor enthusiasts put together literally hundreds of antique tractor shows throughout the United States, Canada, and Europe. While some shows are sponsored by clubs and cater to a certain brand of tractor, others welcome all tractor brands and are open to both steam- and gasoline-powered models. Two of the biggest John Deere shows are the Worldwide Two-Cylinder Expo and The Gathering of the Green. In many cases, you don't even have to take a fully restored model. Go to any one of the shows, and you're sure to see at least one or two tractors setting in the lineup that run just fine, but haven't seen a new coat of paint since they left the factory.

What you will also see are groups of men and women sitting under the closest shade tree, sharing stories, and catching up on each others lives since the last time they all got together. To your benefit, many of them also have tips to share about how they solved a particular problem or located a certain part. You'll find that you're not alone in the challenges you face, as show participants share their war stories.

If you're still in the process of restoring a vintage tractor, a tractor show will often give you the opportunity to closely inspect a like model and ask questions of its owner. Assuming that person has restored his tractor to original condition, there's nothing like physically examining the real thing to know what yours is supposed to look like.

If those aren't reasons enough to participate in a tractor show, consider that many shows feature a swap meet or flea market where you can purchase parts for various models of antique tractors. Some of the shows also feature field demonstrations, tractor parades, and tractor games in which you can participate.

Tractor shows, such as this gathering at the World's Expo of Antique Farm Equipment in Ankeny, Iowa, are ideal places to meet other collectors and show off your accomplishments.

Not all tractors that are on display at a tractor show have been restored. Depending upon their shape, some tractors are actually worth more in their rough condition, provided they run.

Above: The parade of colors and models is a customary part of many tractors shows. Just remember that open platform tractors were designed for only one occupant. Never carry extra passengers on the seat, platform, or, worst of all, the fender.

Right: Tractor shows are a good place to find other models like your own and pick up tips and ideas from other collectors.

Farm Equipment Demonstrations

As the interest in antique tractors has grown, so has the interest in antique farm implements and field demonstrations—and it's easy to see why. When you've got a vintage farm tractor that purrs like a kitten, it's hard to be content just driving it in parades or showing it off to friends. There's always the urge to put it to work in a nostalgic setting.

Hence, many antique tractor shows now feature field demonstrations as part of the agenda. While many started off with plowing demonstrations, the list of activities has grown to include such activities as threshing, baling, shelling corn, cutting silage, and so on. Some, of course, will be stationary demonstrations that use the tractor's belt pulley to power the machine. Others are in the field, where antique tractors can be seen pulling implements that are appropriately matched for the time period and power requirements.

Whether you're watching or participating, it's important to remember one thing: Farm implements built in the early part of this century were not equipped with the safety features, nor the shields and guards, found on today's equipment. Carelessness could cost a finger, arm, or a life at the blink of an eye. So keep your distance from working machines. If you're operating the equipment yourself, or helping a friend, never attempt to make adjustments or clear out a crop slug without first shutting off the tractor. Remember, too, that tractors of the past were never designed for passengers. A tractor fender is not a seat.

Some tractor show organizers include plowing demonstrations in the agenda. Others feature plowing competitions, in which participants are scored on the quality and straightness of the furrow. This one was held at the World's Expo of Antique Farm Equipment in Ankeny, Iowa.

Tractor Games

Gather a bunch of antique tractor enthusiasts together and they're bound to come up with other ways to show off their tractors than just a parade of equipment or a series of field demonstrations. Today, it's not unusual to see tractor shows that list such unusual activities as antique tractor square dances, slow races, and backing contests. All are designed to extend the fun associated with owning and restoring antique tractors.

A slow race, for example, tests not only your tractor-operating skills, but your mechanical skills, as well. The goal of the "race" is to see who can drive a certain distance in the slowest time without stopping and without killing the engine. That means you have to decide which gear to start in, how far you dare throttle the engine back, and how often to apply the brakes. The smoother the engine runs at low speeds, the better you're going to be able to challenge the competition.

If maneuvering a tractor is your forte, you can find plenty of tractor games to test your skills in that area, too. The barrel race, for instance, calls for participants to push a barrel, which has been padded to protect tractor finishes, through an obstacle course. Naturally, drivers with narrow-front tractors tend to hold the advantage.

On the other hand, maybe you'd rather try backing. A couple of contest variations include backing a hay wagon up to a pretend loading dock or through an obstacle course. The winner in either case is the person with the fastest time. Another backing skill game requires participants to see who can back up and stop with the hitch positioned closest to an egg without breaking it. This game is designed to test your skills at hitching up an implement and your ability to line up the hitch pin holes. Occasionally, just to level the playing field, the game organizers require all participants to use the same tractor.

There are other games being played by tractor clubs all over the country. The latest, ever growing in popularity, is tractor square dancing. It takes a lot of room, because it is done a lot like real square dancing. A caller announces the moves to the tune of music, and the tractors drive in a circle and follow the calls.

Antique Tractor Pulls

There are a several reasons antique tractor pulling has become popular with tractor enthusiasts. The first is cost. Compared to modern-day pulling tractors, which are often equipped with multiple turbochargers and high-priced tires, an antique pulling tractor looks pretty much like the original. The only difference is the extra weight racks and the wheelie bars found on some tractors.

Consequently, an antique tractor enthusiast can participate in the sport for a fraction of the cost. On top of that, since the tractors in some classes are not significantly modified, they can still be used for work around the farm or acreage. And most antique tractor pullers simply prefer the slow pace of antique pulls to the glitz, smoke, and noise of modern tractor pulls.

In general, most antique tractor pulls are divided into four or five classes, depending upon the governing body. It used to be that pulling tractors were classified as "antique" if they were built in 1938 or before, while tractors produced from 1939 to 1954 were designated as "classic." The dividing year had more to do with tractor history than anything else. Prior to 1939, things like the radiator, fuel tank, and steering rod were pretty much left exposed, while later-model tractors were more streamlined and typically had more power.

When hauling a tractor—whether it be to a show or parade—make sure the trailer and the towing vehicle are adequate to handle the load and that the tractor is securely attached to the trailer.

Above: Although most antique tractor pulls allow only stock tractors, some include classes for modified tractors, which feature a number of alterations to the original model.

Right: Antique tractor pulls, which often divide vintage models into several weight and model year categories, are a popular way to showcase your mechanical and driving skills.

Today, the National Antique Tractor Pullers Association (NATPA) and United States Antique Pullers (USAP), which generate the rules followed by most sanctioned pulls, have taken both history and tractor size into consideration. They realized that fewer and fewer tractors built prior to 1939 were participating in pulls. The NATPA now has five classes, or divisions, while the USAP recognizes four.

Division I in the NATPA, for example, is designed for beginning pullers and show tractors and is used to promote stock pulling. Only tractors built in 1957 or earlier, or production models that started in 1957, are eligible. In addition, almost everything must be stock and tractors can only pull in low gear with a 2.75 mph speed limit.

Division II is for near-stock tractors that are 1957 or older models. Again, only low gear is permitted, even though some tire modifications are allowed and the speed limit has been increased to 3 mph. By the time you get into Divisions III and IV, drivers are permitted to use any gear, and any kind of cut is allowed on tires. Division III still has a speed limit, while IV does not.

Finally, Division V is for tractors that are 1959 or older. By this stage of the game, any gear and any speed are allowed. So is turning up the engine to 130 percent over manufacturer's data. All divisions except Division I also require wheelie bars.

Meanwhile, tractors competing in USAP events generally have four classes to choose from: Super Farm Stock, Modified Stock, Pro Stock, and Super Pro Stock. Even though there are speed limits in each class, all four classes are limited to tractors that are 1958 or older models.

Keep in mind that there are dozens of tractor pulling organizations around the country, and many of them operate under their own rules and regulations. It's best to check the rules at any pull before you go to the trouble of loading up the tractor and driving any distance.

Just as with modern tractor pulls, antique tractors in each class pull a mechanical sled, on which the weight increases as the sled is pulled down the track. However, there are basically two ways antique tractor pulls are run. One is by weight class and the other is by percentage pull. Weight classes for antique tractors generally start at 3,500 pounds and increase in 500-pound increments, usually up to 7,500 to 8,500 pounds. Determining the winner of each class is just as simple as it is in big pulls—the tractor pulling the sled the farthest wins.

Percentage pulls, on the other hand, require a little bit of math. In this case, you generally want the tractor to be as light as you can get it, or at least at the lighter end of the class. Consequently, any frills added for cosmetic reasons are often left off. Tractors are then weighed to determine their exact weight, and participants attempt to pull as much weight as possible. In the end, if two tractors pull the same amount of weight the same distance, but one weighs 3,500 pounds and the other weighs 4,000 pounds, the lighter tractor would be declared the winner.

The one thing most antique pulls do have in common is the ultimate goal. And surprisingly, it's not the trophy or prize money. What most participants are really after is the pride that comes with winning and the ability to settle the age-old argument—if only for a day—of whether their John Deere can beat their friend's Farmall or Allis-Chalmers.

Appendix
Sources for Parts, Rebuilding, and Repairs

Salvage and Reproduction Parts

Abilene Machine Parts
P.O. Box 129
Abilene, KS 67410
800-332-0239
800-255-0337
785-655-9455
www.abilenemachine.com

Ag Tractor Supply
9301 Breagan Road
Lincoln, NE 68526
800-944-2898
402-421-8822

Aldon Auto Salvage/All Tractor Parts
P.O. Box 3
Lamont, Alberta T0B 2R0
Canada
800-661-8814
780-895-2524
www.aldonauto@aol.com

Antique Tractor Parts
Martin's Farm Supply
5 Ogle Industrial Drive
Vevay, IN 47043
812-427-2622
812-427-4088
www.antiquetractorparts.com

Bill Batt
B & B Tractor
2931 S. Sycamore Grove Road
Garden City, MO 64747
816-682-6888
wbatt@casstel.net

Biewer's Tractor and Restoration Salvage
16242 140th Avenue S.
Barnesville, MN 56514
218-493-4696
bts@rrt.net
www.salvagetractors.com

Central Plains Tractor Parts
712 North Main Avenue
Sioux Falls, SD 57102
800-234-1968
605-334-0021

Colfax Tractor Parts
Rt. 1, Box 119
Colfax, IA 50054
800-284-3001

Dave Cook
28800 Cook Road
Washburn, WI 54891
715-373-2092
Model H parts, restored lights and light service

Correct Connection
3345 Copper Kettle Highway
Rockwood, PA 15557
814-926-2777
Fax: 814-926-2764
corcon@qcol.net
Fasteners and hardware

Dengler's
6687 Shurz Road
Middletown, OH 45042
513-423-4000
Fax: 513-423-0706
www.denglertractor.com

Detwiler Tractor Parts
S3266 Highway 13 S.
Spencer, WI 54479
715-659-4252
Fax: 715-659-3885
www.detwilertractor.com

Dick Moore Repair & Salvage
1540 Joe Quick Road
New Market, AL 35761
205-828-3884

Discount Tractor Supply
P.O. Box 265
Franklin Grove, IL 61031
800-433-5805
815-456-3022

Draper Tractor Parts, Inc.
1951 Draper Road
Garvield, WA 99130
509-397-2666

Dave Geyer
1251 Rohret Road S.W.
Oxford, IA 52322
319-628-4257
JD two-cylinder unstyled tractor hoods; sheet metal fabrication

Fresno Tractor Parts
3444 West Whitesbridge Road
Fresno, CA 93706
209-233-2174

Hart Antique Tractor Reproductions
6333 Highway 45 S.
Paducah, KY 42001
270-554-4028
Fax: 270-554-4955
www.harttractorparts.com
Reproduction grills, hoods, and nose pieces

Hendren's, Inc.
P.O. Box 639
State Road 67 N.
Mooresville, IN 46158
877-216-7120
317-831-1450
www.hendrens.com

Heritage Farm Power, Inc.
P.O. Box 1125
Belton, MO 64012
816-322-1898
www.tractorumbrellas.com
Vintage tractor umbrellas

Klumpp Salvage
P.O. Box 2020
Highway 165 S.
Kinder, LA 70648
800-444-8038
337-738-2554
Fax: 337-738-2456

John R. Lair
205 6th Street W.
Canby, MN 56220
507-223-5902
Fenders

Charles Lindstrom
DBA Lindstrom's JD Model H Parts
1275 NW 26th Road
St. Joseph, MO 64503
816-232-5868
Specializes in Model H parts

Moline Tractor and Plow Co.
320 16th Street
Moline, IL 61265
309-748-7944
Part counter: 866-766-PART (7278)
www.JohnDeereCollectorsCtr.com

Pete's Tractor Salvage, Inc.
Rt. 1, P.O. Box 124
2163 15th Avenue N.E.
Anamoose, ND 58710
800-541-7383
701-465-3274

Dennis Polk Equipment
72435 SE 15
New Paris, IN 46553
800-795-3501
219-831-3555
parts@dennispolk.com
www.dennispolk.com

Renaissance Tractor
120 Cabe Road
Chehalis, WA 98532
800-784-0026
360-748-0026
Specializes in two-cylinder diesels

Restoration Supply Co.
96 Mendon Street
Hopedale, MA 01747
800-809-9156
508-634-6915
www.tractorpart.com

Rock Valley Tractor Parts
1004 10th Avenue
Rock Valley, IA 51247
800-831-8543

Leland Schwandt
14215 486th Avenue
Wilmot, SD 57279
605-432-6192
JD seats and back rests

Sexsmith Used Farm Parts
R. R. 2
Sexsmith, Alberta T0H 3C0
Canada
800-340-1192
780-568-4100

Shepard's 2 Cylinder Parts, Service & Repair
John Shepard
E633-1150 Avenue
Downing, WI 54734
715-265-4988
js2cypts@baldwin-telecom.net

Southeast Tractor Parts
14720 Highway 151
Jefferson, SC 29718
888-658-7171
843-658-7171
setractor@shtc.net

Steiner Tractor Parts, Inc.
10096 S. Saginaw Road
Holly, MI 48442
800-234-3280
810-695-1919
www.steinertractor.com

Taylor Equipment
3694-2 Mile Road
Sears, MI 49679
800-368-3276
231-734-5213
Fax: 231-734-3113

The Tractor Barn
6154 West Highway 60
Brookline, MO 65619
800-383-3678
417-881-3668
www.tractorbarn.net

Tired Iron Farm
19467 County Road 8
Bristol, IN 46507
574-848-4628

2-Cylinder Diesel Shop
Roger and Dana Marlin
731 Farm Valley Road
Conway, MO 65632
417-589-3843

Valu-Bilt Tractor Parts
P.O. Box 3330
Des Moines, IA 50316
888-828-3276
www.valu-bilt.com

Vande Weerd Combine, Inc.
2553 320th Street
Rock Valley, IA 51247
800-831-4814
www.agpartslocator.com (tractor parts)

Wengers of Myerstown
P.O. Box 409
814 South College Street
Myerstown, PA 17067
800-451-5240
717-866-2135
www.wengers.com

Westlock Tractor Parts
P.O. Box 5360
Westlock, Alberta T7P 2P5
Canada
www.westlocktractor.com

Worthington Ag Parts
1923 215th Street
Audobon, IA 50025
www.worthingtonagparts.com

Worthington Ag Parts
2713 N. U.S. 27
St. Johns, MI 48879
800-248-9263

Worthington Ag Parts
Rt. 4, Box 14
Worthington, MN 56187
800-533-5304
www.worthingtonagparts.com

Yesterday's Tractors
P.O. Box 160
Chicacum, WA 98325
www.ytmag.com

To locate additional salvage yards and parts sources in your area, check the National Tractor Parts Dealer Association website at www.NTPDA.com or the ATIS Salvage Yard List at www.atis.net/salyards.shtml.

Carburetors and Governors

Burrey Carburetor Service
18028 Monroeville Road
Monroeville, IN 46773
800-287-7390
219-623-2104

Denny's Carb Shop
8620 N. Casstown-Fletcher Road
Fletcher, OH 45326
937-368-2304
www.dennyscarbshop.com

Link's Carburetor Repair
8708 Floyd Highway N.
P.O. Box 139
Copper Hill, VA 24079
540-929-4519
540-929-4719

McDonald Carb & Ignition
1001 Commerce Road
Jefferson, GA 30549
706-367-9952 (night)
706-367-8851 (7–8 p.m. EST)

Motec Engineering
7342 W. State Road 28
Tipton, IN 46072

Robert's Carburetor Repair
404 E. 5th Street
P.O. Box 624, Dept. GM
Spencer, IA 51301
712-262-5311

Treadwell Carburetor Company
4870 County Highway 14
Treadwell, NY 13846
607-829-8321
www.carbsandkits.com/tc/treadwell

Diesel Injection Pumps and Nozzles
Central Fuel Injection Service Co.
2403 Murray Road
Esterville, IA 51334
712-362-4200
www.centralfuelinjection.com

Roy Ritter
15664 County Road 309
Savannah, MO 64485
816-662-4765
Specializing in two-cylinder diesel injection pumps

Magnetos
The Brillman Company
2328 Pepper Road
Mt. Jackson, VA 22842
888-274-5562
Fax: 540-477-2980
www.brillman.com

Ed Strain
6555 44th Street N., #2006
Pinellas Park, FL 33781
800-266-1623
727-521-1597

Gag Electro Service
Glen Schueler
HCR 2, Box 88
Friona, TX 79035
806-295-3682

Mark's Magneto Service
395 South Burnham Highway
Lisbon, CT 06351
860-887-1094

Larry G. Foster
905 McPherson Road
Burlington, NC 27215
877-556-5421
Fax: 336-584-7563

Lightning Magneto
Rt. 1, County Road 54
Ottertail, MN 56571
218-367-2819

Bill Lopoulos Magneto Parts
304 Pondview Place
Tyngsboro, MA 01879
978-649-7879
www.magnetoparts.com

Magneeders
8215 County Road 118
Carthage, MO 64836
417-358-7863
magneedr@ipa.net

Gauges
Antique Gauges, Inc.
12287 Old Skipton Road
Cordova, MD 21625
410-822-4963

Wiring Harnesses
Agri-Services
13899 North Road
Alden, NY 14004
716-937-6618
www.wiringharnesses.com

Seals and Gaskets
A-1 Leather Cup and Gasket Company
2103 Brennan Circle
Fort Worth, TX
817-626-9664

Lubbock Gasket & Supply
402 19th Street, Dept. AP
Lubbock, TX 79401
800-527-2064
806-763-2801

Olson's Gaskets
3059 Opdal Road E.
Port Orchard, WA 98366
www.olsonsgaskets.com
360-871-1207

Radiator (Repair)
Omaha Avenue Radiator Service
100 E. Omaha Avenue
Norfolk, NE 68701
402-371-5953
eclkcl@kdsi.net

Sieren's Reproduction Radiator Shutters
1320 Highway 92
Keota, IA 52248
319-698-4042

Wheels and Rims
Detwiler Tractor Parts
S3266 Highway 13 S.
Spenser, WI 54479
715-659-4252
Fax: 715-659-3885
www.detwilertractor.com

Nielsen Spoke Wheel Repair
Herb Nielsen
3921 230th Street
Estherville, IA 51334
712-867-4796

TNT Poly Div.
Taube Toll Corp.
1524 Chester Boulevard
Richmond, IN 47374
765-962-7415
TNTPoly@aol.com
Polyurethane replacement lugs for steel wheels

Wilson Farms
20552 Old Mansfield Road
Fredricktown, OH 43019
740-694-5071

Tires
Gempler's, Inc.
100 Countryside Drive
P.O. Box 270
Belleville, WI 53508
800-332-6744
Order line: 800-382-8473
www.gemplers.com

M. E. Miller Tire Co.
17386 State Highway 2
Wauseon, OH 43567
419-335-7010
www.millertire.com

Tucker's Tire
844 S. Main Street
Dyersburg, TN 38024
800-443-0802

Wallace W. Wade Specialty Tires
P.O. Box 560906
Dallas, TX 75356
214-688-0091
800-666-TYRE
www.wallacewade.com

Replacement Seats
Speer Cushion Co.
431 S. Intercean Avenue
Holyoke, CO 80734
800-525-8156
www.speercushion.com

Mufflers
Oren Schmidt
2059 V Avenue
Homestead, IA 52236
319-662-4388

Jim Van DeWynckel
R. R. 4
Merlin, Ontario N0P 1W0
Canada

Steering Wheels (Recovering)
Tom Lein
24185 Denmark Avenue
Farmington, MN 55024
651-463-2141

Minn-Kota Repair
RR 1, Box 243
Ortonville, MN 56278
320-839-3940
320-289-2473

Tractor Steering Wheel Recovering
 and Repair
1400 121st Street W.
Rosemount, MN 55068
612-455-1802

Decals
Jorde's Decals
935 Ninth Avenue N.E.
Rochester, MN 55906
507-288-5483
decals@jordedecals.com
www.jordedecals.com

Kenneth Funfsinn
Rt. 2
Mendota, IL 61342

K & K Antique Tractors
5995 N. 100 W.
Shelbyville, IN 46176
317-398-9883
www.kkantiquetractors.com

Jack Maple
Rt. 1, Box 154
Rushville, IN 46173
317-932-2027
Decals for a wide variety of applications and models

Dan Shima
409 Sheridan Drive
Eldridge, IA 52748
319-285-9407

Restoration Equipment and Supplies

E & K Ag Products
HCR 3, Box 905
Gainesville, MO 65655
417-679-3530
ekag@webbound.com
Sleeve puller

CJ Spray, Inc.
370 Airport Road
South St. Paul, MN 55075
1-800-328-4827
Spray systems

Jim Deardorff
P.O. Box 317
Chillicothe, MO 64601
660-646-6355
Fax: 660-646-3329
jdeardorff@yahoo.com
Classic Blast sandblasting mix (made from aluminum oxide and black walnut shells)

TP Tools and Equipment
Dept. AP, 7075 Rt. 446,
P.O. Box 649
Canfield, OH 44406
800-321-9260
Info Line: 330-533-3384
www.tiptools.com
Parts washers, grinders, presses, sandblasting equipment, etc.

Publications and Clubs

Tractor Manuals
Jensales Inc.
P.O. Box 277
Clarks Grove, MN 56016
800-443-0625 (orders)
507-826-3666
www.jensales.com

Clarence L. Goodburn Literature Sales
101 W. Main
Madelia, MN 56062
507-642-3281

King's Books
P.O. Box 86
Radnor, OH 43066

Yesterday's Tractors
P.O. Box 160
Chicacum, WA 98325
www.ytmag.com

Intertec Publishing
P.O. Box 12901
Overland Park, KS 66282
800-262-1954
www.intertecbooks.com

General Magazines
Antique Power
P.O. Box 500
Missouri City, TX 77459

The Belt Pulley
20114 IL Rt. 16
Nokomis, IL 62075

The Hook Magazine
P.O. Box 16
Marshfield, MO 65706
417-468-7000
Tractor pulling, including antique and classic

John Deere Newsletters/Magazines
Green Magazine
Dept. SG
2652 Davey Road
Bee, NE 68314
402-643-6269
www.greenmagazineonline.com

Two Cylinder Club/Publications
P.O. Box 10
Grundy Center, Iowa 50638-0010
www.two-cylinder.com

John Deere Tradition
1503 S.W. 42nd Street
Topeka, KS 66609
800-678-4883

John Deere Historical Records
John Deere Collectors Center
320 Sixteenth Street
Moline, IL 61265
866-748-7944
www.JohnDeereCollectorsCtr.com

Two Cylinder Club
P.O. Box 430
Grundy Center, IA 50638
888-782-2582
319-345-6060

Index

axles, front, *110–115*
    inspection, *113–114*
    rebuilding, *113–114*
axles, rear, *103, 106–107, 197*
    inspection, *106–107*
    seal repair, *106–107*
belts, *171*
body work, *172–181*
    dent and crease repair, *176*
    hole and rust repair, *176–179*
    sheet metal repair, *176–181*
    surface preparation, *178–180*
bolts,
    easy-outs, *52*
    removing broken or damaged, *52*
brakes, *102, 108*
    adjustment, *109*
    inspection, *108–109*
    rebuilding, *108–109*
cam followers, *78*
camshafts, *70, 77–78, 87*
    bearings/bushings, *77*
    inspection, *70, 78*
carburetors, *151–155*
    adjustment, *151*
    checking float bowl level, *151*
    disassembly, *153–154*
    inspection, *155*
    rebuilding, *153–156*
clutches, *95, 97–101*
    adjustment, *97, 99, 100–101*
    foot-clutch systems, *99*
    hand-clutch systems, *99–101*
    inspection, *97*
    throw-out bearing, *99*
coils, *140*
connecting rod bearings, *74–76*
connecting rods, *70, 74–76*
cooling system, *162–171*
    hoses, *168*
    inspection, *55, 162–164*
    kerosene engines, *165*
    operation, *162, 165*
    rebuilding, *163–171*

crankshafts, *74, 75, 87*
    inspection, *74*
    grinding, *74*
    scoring repair, *74*
cylinder heads, *64, 89–90*
cylinders, *58–59, 64, 69*
    boring, *86*
    honing, *69, 72, 87*
    ridge reaming, *64*
    sleeves installation, *87*
    wall scoring, *64*
decals, *198–205*
    application, *202–205*
    researching originality and placement, *199–201*
    surface preparation, *202*
diesel injection pumps, *156–157*
diesel systems, operation, *156*
differentials, *103–105*
    inspection, *105*
    rebuilding, *105*
distributors, *138–140*
    inspection, *138–140*
    repair, *138–140*
electrical system, *57, 132–147*
    inspection, *56*
emblems, restoration, *206*
engine blocks, *64, 66, 73*
    mating surface flatness, *64*
    preparation, *64*
    welding cracks, *64*
engines, *56, 57, 60–93*
    compression testing, *58–59*
    disassembly, *64*
    freeing a stuck engine, *62–63*
    inspection, *55–57, 64*
    rebuild kits, *87*
    rebuilding, *87–91*
    troubleshooting, *54–59*
    excessive exhaust smoke, *56, 57, 60, 64*
    hard starting, *56*
    knocking, *60*
    loss of power, *84, 100*

    overheating, *57, 100*
    uneven running, *56, 57, 58*
fans, *166–167*
final drives, *102–105*
    gear mesh adjustment, *105*
    inspection, *105*
    rebuilding, *105*
fuel hoses, inspection, *151*
fuel system, *148–161*
fuel tanks, *148–150*
    cleaning, *148–149*
    repairing leaks, *149–150*
    sealing, *150*
gauges, restoration, *147*
generators, *140–142*
    inspection, *141–142*
    operation, *140–142*
    rebuilding, *141–142*
governors, *160*
    adjustment, *160*
    inspection, *160–161*
    overhaul, *160*
historical records, *33–34*
hydraulic system, *126–131*
    contamination, *129*
    inspection, *56–57, 130*
    operation, *126–129*
    rebuilding, *130–131*
    seals, *131*
    troubleshooting, *130*
lights, *146*
magnetos, *132–137*
    inspection, *134–136*
    operation, *133*
    service, *136*
    timing, *136–137*
main bearings, measuring, *75*
manifolds, *91, 159*
    inspection, *159*
    repair, *159*
name plate restoration, *206*
oil leaks, *54, 56, 92, 106*
oil pans, *92*

oil pumps, *92*
    inspection, *92*
    rebuilding, *92*
oil-bath air filters, *57, 158*
paint types, *185–186*
    acrylic enamel paints, *185–186*
    lacquer paints, *185*
    urethane pants, *186*
painting, *182–197*
    applying paint, *188–196*
    ensuring paint compatibility, *186*
    equipment, *187*
    matching paint colors, *185*
    paint removal, *173–175*
    preparation, *173–175, 182*
    primer coat, *182–184*
parts fabrication, *43, 176*
piston pins, *86*
    inspection, *86*
    pin-to-bushing tolerances, *86, 87*
piston rings, *64, 69–73*
    determining replacement size, *69, 72*
    end gap adjustment, *69–70, 72*
    grooves inspection, *69, 71*
    installation, *70, 72, 87*
    removal, *69, 70*
    replacement, *69–72*
piston sleeves, *64, 86*
pistons, *64, 73, 86*
    direction of orientation, *64, 65*
    inspection, *64, 86*
    installation, *87*
    piston-to-cylinder wall measurements, *86, 87*
    removal, *64*
    scoring, *64, 86*
pony engines, *93*
power steering, *115–117*
    rebuilding, *116–117*
power takeoff (PTO), *101*
    inspection, *101*
    operation, *101*
    repair, *101*
primers, *182–184*
    epoxy primers, *184*
    filler primers, *183*
    sealing primers, *183*
    urethane primers, *184*

push rods, *77, 78*
radiator caps, *57, 166*
replacement parts, *172–173*
    locating, *50–51*
    reproduction, *50–51*
rocker arms, inspection, *78*
rod-bearing caps, *70, 76*
    removal, *64*
rod bearings, measuring, *74–75*
Roll-O-Matic, rebuilding, *115*
serial number plates, restoration, *206–207*
shafts, repair, *53*
sheet metal, see body work
spark plug wires, *146*
Speedi-Sleeves, *53*
Starters, *57, 143*
    inspection, *143*
    rebuilding, *143*
steering wheels, restoration, *117*
tappets, adjustment, *64, 78, 80, 85*
thermostats, *171*
tires, repair, *120–121*
tools,
    shopping for, *40*
    specialized, *42–45*
tractor disassembly, *49–50*
    cleaning, *48–49*
    paint removal, *173–175*
tractor evaluation, *55–57*
    cooling system, *55*
    electrical system, *56, 57*
    engine noise, *56*
    engine running, *56, 57*
    engine starting, *56*
    exhaust, *56*
    hydraulic fluid, *56*
    hydraulic system, *56–57*
    oil leakage, *56*
    oil quality, *56*
    transmission fluid, *56*
transmission bearings, inspection and replacement, *96, 101*
transmission seals, inspection and replacement, *96, 101*
transmissions, *95–97*
    inspection, *95–97*
    troubleshooting, *95*
    repair, *95–97*

valve guides, *64, 78, 80, 82, 83*
valve keepers, *78, 81, 85*
valve seats, *78, 80, 81, 82, 84*
valve springs, *80, 83, 84, 85*
    inspection, *80*
    removal, *80–81*
    tension, *80*
valves, *78–85*
    inspection, *78, 79*
    grinding, *78, 80, 81, 83*
    lapping, *78, 80, 81, 83–85*
    removal, *80–81*
valvetrains, overhaul, *80–85*
voltage regulators, operation, *140–142*
water pumps, *169–170*
    inspection, *169*
    rebuilding, *169*
wheels and rims, rebuilding, *121–125*
wiring, *144–145*
    inspection, *144*
    restoration, *144–145*

About the Author

Tharran E. Gaines was born in north-central Kansas where he grew up on a small grain and livestock farm near the town of Kensington. The only boy in a family of five children, he still feels that he did all of the farm work normally delegated to a full farm family of boys; others say he was spoiled by four younger sisters.

He attended Kansas State University, where he received a degree in wildlife conservation and journalism with the goal of going into outdoor writing. He soon was using his agricultural background as a technical writer for Hesston Corporation and hasn't left agriculture since.

As a technical writer, he has produced repair manuals, owner's manuals, and assembly instructions for Hesston, Winnebago, Sundrstrand hydrostatic transmissions, Kinze planters and grain wagons, and Best Way crop sprayers. As a creative writer, he has crafted and produced everything from newsletter and feature articles to radio and TV commercials to video scripts and advertising copy for such companies as DeKalb Seed Company, DowElanco, Asgrow Seed Company, Rhone-Poulenc, Farmland Industries, and AGCO Corporation.

In 1991, he began his own business as a freelance writer and today continues to operate Gaines Communications with his wife, Barb, out of their home office. The majority of their business involves producing all editorial copy for two AGCO Corporation company magazines under contract to *Progressive Farmer*. These magazines include AGCO *Advantage* and Hesston *Prime Line*.

Tharran and his wife have two grown children, Michael and Michelle. Both have moved away, however, leaving just the two of them to share their 100-plus-year-old Victorian home in Savannah, Missouri.